AUTONOMIC
NETWORKING-ON-CHIP

Bio-Inspired Specification, Development, and Verification

Embedded Multi-Core Systems

Series Editors

Fayez Gebali and Haytham El Miligi
University of Victoria
Victoria, British Columbia

Autonomic Networking-on-Chip: Bio-Inspired Specification, Development, and Verification, *edited by Phan Cong-Vinh*

Bioinformatics: High Performance Parallel Computer Architectures, *edited by Bertil Schmidt*

Multi-Core Embedded Systems, *Georgios Kornaros*

AUTONOMIC
NETWORKING-ON-CHIP

Bio-Inspired Specification, Development, and Verification

Edited by Phan Cong-Vinh

CRC Press
Taylor & Francis Group
Boca Raton   London   New York

CRC Press is an imprint of the
Taylor & Francis Group, an **informa** business

CRC Press
Taylor & Francis Group
6000 Broken Sound Parkway NW, Suite 300
Boca Raton, FL 33487-2742

First issued in paperback 2017

© 2012 by Taylor & Francis Group, LLC
CRC Press is an imprint of Taylor & Francis Group, an Informa business

No claim to original U.S. Government works

ISBN-13: 978-1-4398-2911-0 (hbk)
ISBN-13: 978-1-138-07673-0 (pbk)

Contents

List of Figures

List of Tables

Foreword

Systems on a chip are complex embedded systems focusing less on computation and increasingly on communication. This shifts design style based on platforms (i.e., design templates) to communication-based design. In this new paradigm, network-on-a-chip has emerged as a new type of interconnect that can solve this problem. Furthermore, for adaptive communication, autonomic networking-on-chip (ANoC) has widely attracted attention from researchers in the network-on-chip field due to its unique advantages and has become one mainstream in the network-on-chip society. Unfortunately, there is no book specifically focused on formal aspects of ANoC. But there has been a strong research need worldwide to strengthen information exchange in the area of ANoC. This edited book is one of salient documents for disseminating research results related to the formalization of ANoC.

Autonomic Networking-on-Chip: Bio-Inspired Specification, Development and Verification contains original, peer-reviewed chapters reporting on new developments of interest to both the autonomic computing and network-on-chip communities in all remarkable topics of ANoC from specification to implementation for ANoC based on rigorous interdisciplinary approaches in which theoretical contributions have been formally stated and justified, and practical applications have been based on their firm formal basis. In other words, the major technical contents of the book include the following: bio-inspired NoC, mapping of applications onto ANoC, ANoC for FP-GAs and structured ASICs, applications of formal methods in ANoC development, formalizing languages that enable ANoC, validation and verification techniques for ANoC, self-* in ANoC (such as self-organization, self-configuration, self-healing, self-optimization, self-protection and so on), calculi for reasoning about context awareness in ANoC and programming models for ANoC.

The book has come into being from a fine collection of chapters emphasizing the multidisciplinary character of investigations from the point of view of not only the ANoC field involved but also the formal methods. The book, which is specially dedicated to reporting on recent progress made in the field of bio-inspired on-chip networks (BioChipNets), will usefully serve as a technical guide and reference material for engineers, scientists, practitioners, and researchers by providing them with state-of-the-art research results and future opportunities and trends. To the best of my knowledge, this is the first book that presents achievements and findings of ANoC research covering the full spectrum of formalizing BioChipNets. These make the book unique and, in more than one respect, a truly valuable source of information that may be considered a landmark in the progress of ANoC. Congratulations to those who have contributed to the highest technical quality of the book!

Petre Dini
Advisory Committees Board Chair of IARIA

Preface

A new networking-on-chip paradigm, which is seen as a cutting-edge approach to network-on-chip (NoC), is currently on the spot as one of the priority research areas: autonomic networking-on-chip (ANoC), which is inspired by the human autonomic nervous network.

ANoC is a networking-on-chip paradigm able to realize its self-* functionality such as self-organization, self-healing, self-configuration, self-optimization, self-protection, and so on whose context awareness is used to control networking functions dynamically. The overarching goal of ANoC is to realize intelligent networks-on-chips that can manage themselves without direct human interventions. Meeting this grand challenge of ANoC requires a rigorous approach to ANoC and the notion of self-*. To this end, taking advantage of formal engineering methods we will establish, in this book, formal and practical aspects of ANoC through specifying, refining, programming and verifying ANoC and its self-*. All of these are to achieve foundations and practice of ANoC.

From the above characteristics, novel approaches of specification, refinement, programming and verification are arising in formal engineering methods for ANoC. Therefore, new methodologies, programming models, tools and techniques are imperative to deal with the impact of ANoC and its self-* mentioned above on emerging intelligent networks-on-chips.

The book is a reference for readers who already have a basic understanding of NoC and are now ready to know how to bio-inspiredly specify, develop and verify ANoC using rigorous approaches. Hence, the book includes both theoretical contributions and reports on applications. To keep a reasonable tradeoff between theoretical and practical issues, a careful selection of the chapters was completed, on one hand, to cover a broad spectrum of formal and practical aspects and, on the other hand, to achieve as much as possible a self-contained book.

Formal and practical aspects will be presented in a straightforward fashion by discussing in detail the necessary components and briefly touching on the more advanced components. Therefore, bio-inspired specification, development and verification demonstrating how to use the formal engineering methods for ANoC will be described using sound judgment and reasonable justifications.

This book, with chapters contributed by prominent researchers from academia and industry, will serve as a technical guide and reference material for engineers, scientists, practitioners and researchers by providing them with state-of-the-art research findings and future opportunities and trends. These contributions include state-of-the-art architectures, protocols, technologies, and applications in ANoC. In particular, the book covers existing and emerging research issues in bio-inspired on-chip networks (BioChipNets). The book has eight chapters addressing various topics from specification to implementation of ANoC based on rigorous interdisciplinary approaches.

Chapter 1 by M. Bakhouya presents an overview of state-of-the-art approaches for ANoC. Unlike design-time approaches in which all parameters/protocols are optimized/selected at design time targeting a specific application, in run-time approaches the adaptation process is running continuously to evolve the system. Based on this state-of-the-art review, the self-* capabilities are considered in the context of ANoC. These capabilities can provide the core scheme to develop ANoC at application, communication and architecture levels. An approach inspired by a biological immune system toward developing ANoC with these self-* capabilities is introduced. One aspect of the self-healing capability is partially developed, and some preliminary results are reported and showed that the adaptation is energy and latency efficient.

Chapter 2 by A. A. Morgan, H. Elmiligi, M. W. El-Kharashi, and F. Gebali[1] discusses a GA-based multi-objective technique for autonomous on-chip network architecture optimization. This technique considers four NoC metrics: power, area, delay, and reliability. The models of the four metrics are discussed. The formulation of the fitness function is then presented. Finally, the GA representation of the problem is explained. The optimization can be carried out for power, area, delay, reliability or the four of them according to weight factors supplied by the designer. Results show that the technique is an efficient way to compromise between different NoC metrics. Moreover, the running time of the technique makes it more suitable for ANoC than previous architecture optimization techniques.

Chapter 3 by K. Latif, A. M. Rahmani, T. Seceleanu, and H. Tenhunen discusses an autonomous PVS-NoC architecture where an ideal tradeoff among virtual channel utilization, system performance and power consumption is presented. The virtual channel buffers are shared between two input ports in PVS-NoC. Apart from system performance, the architecture is also fault tolerant. In case of any link failure, the VC buffers are used by the other physical channel sharing the VC buffers, and system performance is not affected severely. The architecture is simulated with synthetic and real application traffic patterns. The performance is compared with typical VC-based NoC architecture and FVS-NoC. The PVS-NoC architecture shows significant improvement in system throughput without significant power consumption overhead.

Chapter 4 by J. Jaros and V. Dvorak discusses an evolutionary design that is able to produce optimal or near optimal communication schedules comparable to or even better than those obtained by a conventional design for the networks sizes of interest. Moreover, evolutionary design reduces many drawbacks of present techniques

---

[1]MATLAB®, the language of technical computing mentioned in Chapter 2, is a registered trademark of The MathWorks, Inc. For product information, please contact:
The MathWorks, Inc.
3 Apple Hill Drive
Natick, MA 01760-2098 USA
Tel: 508 647 7000
Fax: 508-647-7001
E-mail: info@mathworks.com
Web: www.mathworks.com

and invents still unknown schedules for an arbitrary topology and scatter/broadcast communication patterns.

Chapter 5 by P. C. Vinh breaks new ground in dealing with the *core-to-core* and *agent-based networking techniques* for ANoC using a categorical approach of tasks and data parallel processing – the firm formal method applicable to a wide variety of BioChipNets. The major contribution of the chapter is to propose some applied categorical structures of tasks parallel and data parallel for parallel processing on BioChipNets that has never been tackled thoroughly in this emerging field. By this approach, category theory is applied to deal in an abstract way with algebraic objects and relationships between them for specifying tasks parallel and data parallel on BioChipNets. The chapter shows that for specifying, analyzing and verifying tasks and data parallelism, the categorical approach becomes much better approaching than other ones in theory of algebras. From an applicative aspect of the approach, moreover, formalizing tasks parallel and data parallel on BioChipNets is a categorical specification of middleware that can be used to develop implementations for core-to-core and agent-based networking.

Chapter 6 by L. Guang, J. Plosila, J. Isoaho, and H. Tenhunen presents a novel design approach, hierarchical agent-monitored SoC (HAMSoC), by elaborating its formal specification framework and demonstrating a design example of hierarchical power monitoring on NoCs. This approach provides scalability in terms of both design effort and physical overhead. The formal specification framework of HAMSoC enables efficient system-level specification and modeling for early-stage development. It adopts a highly abstracted formal language to specify exclusively the monitoring operations and parameters, with proper exposure of SW/HW interfaces. The FSM model is used for state transition analysis in monitoring operations. The specification is transformed into generic formal languages so that existing verification tools can be used for further system development.

Chapter 7 by L. Petre et al. discusses a middleware language based on formal methods that has been developed for real networks to be adapted to applications running on NoC. The chapter presents how to place modules that communicate often in each other's vicinity for efficiency. This becomes even more interesting if the NoC consists of vertically stacked layers, commonly referred to as 3-D NoCs. Furthermore, three dynamic alternatives for replacing an application are analyzed, some of which are centralized and others are distributed.

Finally, Chapter 8 by H. Zakaria, E. Yahya, and L. Fesquet discusses a survey on different problems facing designers over the nanometric era. Analyses and solutions to each of these challenges are presented that enable the design of a self-adaptable SoC. GALS systems are considered and some proposed solutions are also applicable to simpler designs.

This title is the third book in a series on Embedded Multi-Core Systems (EMS) published by CRC Press. The EMS series is under the editorial supervision of the Integrated Microsystems Research Group, which is one of the research groups of the Electrical and Computer Engineering Department (ECE), Faculty of Engineering at the University of Victoria (UVic) in Canada.

This book has the following remarkable features:

- Provides a comprehensive reference on ANoC

- Presents state-of-the-art techniques in ANoC

- Formally specifies, develops and verifies bio-inspired on-chip networks

- Includes illustrative figures facilitating easy reading

- Discusses emerging trends and open research problems in ANoC

We owe our deepest gratitude to Professor Fayez Gebali (ECE, UVic) and Dr. Nguyen Manh Hung (NTT University) for their useful support, notably to all the authors for their valuable contributions to this book and their great efforts. All of them are extremely professional and cooperative. We wish to express our thanks to CRC Press, especially Nora Konopka, Kari Budyk, Stephanie Morkert and Karen Simon for their support and guidance during the preparation of this book. A special thank you also goes to our families and friends for their constant encouragement, patience, and understanding throughout this project.

The book serves as a comprehensive and essential reference on autonomic networking-on-chip and is intended as a textbook for senior undergraduate- and graduate-level courses. It can also be used as a supplementary textbook for undergraduate courses. The book is a useful resource for students and researchers to learn autonomic networking-on-chip. In addition, it will be valuable to professionals from both academia and industry and generally has instant appeal to people who would like to contribute to autonomic networking-on-chip technologies.

We highly welcome and greatly appreciate your feedback and hope you enjoy reading the book.

Phan Cong-Vinh
Ho Chi Minh City, Vietnam

About the Editor

Phan Cong-Vinh received a PhD in computer science from London South Bank University (LSBU) in the United Kingdom, a BS in mathematics and an MS in computer science from Vietnam National University (VNU) in Ho Chi Minh City, and a BA in English from Hanoi University of Foreign Languages Studies in Vietnam. He finished his PhD dissertation with the title *Formal Aspects of Dynamic Reconfigurability in Reconfigurable Computing Systems* supervised by Prof. Jonathan P. Bowen at LSBU where he was affiliated with the Centre for Applied Formal Methods (CAFM) at the Institute for Computing Research (ICR). From 1983 to 2000, he was a lecturer in mathematics and computer science at VNU, Posts and Telecommunications Institute of Technology (PTIT) and several other universities in Vietnam before he joined research with Dr. Tomasz Janowski at the International Institute for Software Technology (IIST) in Macao SAR, China, as a fellow in 2000. From 2001 to 2010 he did research together with Prof. Jonathan P. Bowen as a research scholar and then as a collaborative research scientist at CAFM. From January 2011 to May 2011 he worked for the FPT - Greenwich collaborative program at FPT University (FU) in Vietnam as a visiting lecturer. From June 2011 to the present he has become a member of NTT University (NTTU) to take on the responsibilities of an IT Department's deputy dean. He has been author or co-author of many refereed contributions published in prestigious journals, conference proceedings or edited books. He is the author of a book on computing science titled *Dynamic Reconfigurability in Reconfigurable Computing Systems: Formal Aspects of Computing* (2009) and editor of two titles in addition to the present work, *Formal and Practical Aspects of Autonomic Computing and Networking: Specification, Development and Verification* (IGI Global) and *Advances in Autonomic Computing: Formal Engineering Methods for Nature-Inspired Computing Systems* (Springer), to be published in 2012. He is also an IEEE member. His research interests center on all aspects of formal methods, autonomic computing and networking, reconfigurable computing, ubiquitous computing, and applied categorical structures in computer science.

List of Contributors

M. Bakhouya
Universite de Technologie de Belfort Montbeliard, France

V. Dvorak
Brno University of Technology, Czech Republic

M. W. El-Kharashi
Ain Shams University, Egypt

H. Elmiligi
University of Victoria, Canada

L. Fesquet
TIMA CNRS, Grenoble Institute of Technology, UJF, France

F. Gebali
University of Victoria, Canada

L. Guang
University of Turku, Finland

J. Isoaho
University of Turku, Finland

J. Jaros
Brno University of Technology, Czech Republic

K. Latif
University of Turku, Finland
Turku Centre for Computer Science (TUCS), Finland

P. Liljeberg
University of Turku, Finland

A. A. Morgan
University of Victoria, Canada

L. Petre
Åbo Akademi University, Finland

J. Plosila
University of Turku, Finland

A. M. Rahmani
University of Turku, Finland
Turku Centre for Computer Science (TUCS), Finland

T. Seceleanu
ABB Corporate Research, Västerås, Sweden
Mälardalen University, Västerås, Sweden

K. Sere
Åbo Akademi University, Finland

H. Tenhunen
University of Turku, Finland
Turku Centre for Computer Science (TUCS), Finland

L. Tsiopoulos
Åbo Akademi University, Finland

P. C. Vinh
NTT University, Vietnam

E. Yahya
TIMA CNRS, Grenoble Institute of Technology, UJF, France
Banha High Institute of Technology, Egypt

H. Zakaria
TIMA CNRS, Grenoble Institute of Technology, UJF, France
Banha High Institute of Technology, Egypt

1

A Bio-Inspired Architecture for Autonomic Network-on-Chip

M. Bakhouya

Universite de Technologie de Belfort Montbeliard, Belfort, France

CONTENTS

Network-on-chip has been recently proposed for SoC (System-on-Chip) applications design to achieve better performance and lower energy consumption when compared with conventional on-chip bus architectures. The past few years' research in the domain of NoC (Network-on-Chip) has been concentrated on application-specific approaches. These approaches are design-time parameterized and do not consider runtime configuration of different NoC parameters, which are hard to predict in early development stages. More precisely, architecture and system parameters, such as routing algorithm, switching scheme and flits' size, should be adapted at runtime. To address the uncertainty in NoC applications and efficiently use resources, NoC architectures must be adaptive, for example, by rerouting at run time traffic from congested area or by dynamically changing links bandwidth. In this chapter, run-time approaches proposed for autonomic NoCs (ANoCs) are first presented. An approach inspired by the immune system for ANoCs is introduced. The immune system has a useful set of organizing principles, such as self-configuration, self-optimization, and self-healing, that can guide the design of autonomic network-on-chip.

1.1   Introduction

The effectiveness of SoCs is often determined by on-chip interconnect (OCI). SoC applications require the OCI to handle a large amount of traffic by providing low latency communications and high throughput while minimizing the area overhead and energy consumption. The interconnect architectures used in current SoCs to integrate components are based on bus schemes. With the increasing complexity of SoC and its communications requirements, NoC has emerged as the pervasive communication fabric to connect different IP (Intellectual Property) core elements in many-core chips and as a solution of nonscalable shared bus schemes [72, 291]. The main objectives are to satisfy quality of service requirements, to optimize resources' usage, and to minimize the energy consumption by separating the communication from the computation.

Different NoC based architectures using packet-switching have been studied and adapted recently for SoC design. Examples of these architectures are Fat-Tree (FT), 2D mesh, Ring, Butterfly-Fat Tree (BFT), Torus, Spidergon, Octagon, and WK [237, 290]. These on-chip interconnect architectures draw on concepts inherited from parallel and general computer networks, which are well-established research topics. There are differences between these two systems, despite some issues they have in common. For example, in NoCs, switches are more resource-constrained (e.g., silicon area) than those in general computer networks. Communication links of NoCs are relatively shorter than those in computer networks, allowing tight synchronization between routers. All these difference make the tradeoffs in their design very different [72]. Therefore, NoCs should be scalable and adaptive to support various applications by selecting the most suitable parameters based on the requirements of the current application and system conditions.

Several approaches have been proposed to deal with NoCs design and can be classified into two main categories, design-time approaches and run-time approaches. Design-time approaches are generally tailored an application domain or a specific application by providing an application-specific NoC. All parameters, such as the on-chip interconnect architecture (i.e., topology), routing, and switching schemes, are defined at design time [29, 234]. Run-time approaches, however, provide techniques that allow a NoC to autonomously adapt its structure and its behavior during the course of its operation (i.e., at run time).

Recently, there has been a great deal of interest in the development of run-time approaches for autonomic NoC. These approaches can be classified, as illustrated in Figure 1.1, based on the level in which the adaptation is carried out, at the application level, at the communication level, or at the architecture level. In this chapter, a state of the art review of run-time approaches proposed in the literature for ANoCs with adaptive capabilities are presented. Adaptive capabilities have been seen in natural and biological systems and have inspired many researches to develop adaptive systems. In other words, biological and natural systems have been exploited in a variety of computation systems and been perceived as an efficient system model for developing adaptive systems [10, 102, 204] and reconfigurable and evolvable hardware systems [112, 270, 295, 303].

After describing run-time approaches, a bio-inspired architecture, called BNoC, inspired by biological immune system (BIS) is introduced for autonomic NOCs. Biological immune system could allow the design of autonomic NoCs with promising adaptive capabilities. The objective is to implement these capabilities within the system to adapt to environment changes and the dynamic of its computing elements. BNoC could react like an immune system against pathogens that have entered the body. It detects the infection (i.e., applications behavior or system state changes) and delivers a response to eliminate it (i.e., adapt to changes). The aim is to highlight how biological immune system principles can be incorporated into the design of autonomic middleware for NoCs.

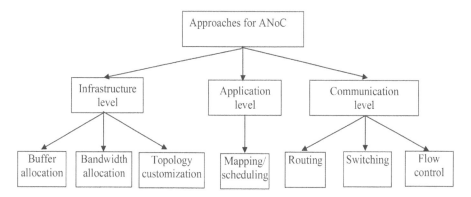

FIGURE 1.1
Classification of run-time approaches for ANoCs.

The rest of this chapter is organized as follows. In section 2, we present run-time approaches proposed for adapting the application to the NoC architecture. Section 3 presents approaches dealing with the adaptation at the communication level. In section 4, we present approaches proposed mainly to optimize some architecture parameters. In section 5, approaches dealing with mapping and scheduling issues at the application level are presented. A bio-inspired approach is introduced and some preliminary results are presented in section 6. Conclusions and future work are given in section 7.

1.2   Infrastructure level

At the infrastructure layer, the performance and the efficiency of the NoC are highly dependent on the on-chip interconnect. Switches constitute the active component that influences on latency and throughput. For example, designing efficient switches with

minimum buffer size (a critical resource) represents a main issue for the success of
NoC design. Links are components that transport data between switches, efficient
allocation of their capacity, decreasing their length or increasing their number may
increase the performance of SoC applications. Hence, on-chip interconnect configuration, bandwidth allocation, and buffer minimization are three main issues that
should be addressed when designing ANoC. The rest of this section describes approaches mainly proposed to address these issues.

1.2.1   Topology customization

Many studies have shown that to improve the performance of specific application
domain the NoC architecture must be customized by inserting additional links between switches. More precisely, on-chip interconnect architecture can be designed
and customized for a priori known set of computing resources (IP cores), and given
pre-characterized traffic patterns among them [39]. These approaches can be classified into two main classes: synthesizing approaches and customizing approaches.
The first class concerns techniques proposed to fully customize the architecture that
is more suitable for a particular application and does not necessarily conform to regular topologies (e.g., 2D mesh). These techniques, such as presented in [221,246,284],
despite their capabilities to maximize the performance by offering techniques for incorporating synthesized on-chip interconnect architectures that are more suitable for
a given application, they suffer from providing architectures with high regularity in
reasonable time [194]. Customizing approaches, however, focus mainly to start with
a regular on-chip interconnect architecture like 2D mesh and partially customize it
by inserting long-range links. For example, an algorithm for customizing a standard mesh NoC by inserting few application-specific long-range links was proposed
in [234,289].

These approaches are mainly proposed to customize/synthesize, at design time,
on-chip interconnects given an application traffic workload. They deal with the
selection of an on-chip interconnect architecture to accommodate the expected
application-specific data traffic patterns during design space exploration phase. For
autonomic NoC, in which traffic pattern is not known or predictable in advance, run-
time approaches are required.

Recently, approaches to dynamically reconfigure the on-chip interconnect by establishing at runtime links are proposed. For example, a power-aware network approach has been proposed in [280]. This approach responds to the bursts and dips in
traffic by turning links "on" and "off" at runtime. However, this approach is based on
the assumption that the future traffic patterns are predictable based on the observation
of the past traffic.

In [308], an approach is proposed to dynamically establish interconnections by
adapting the physical network. Similar to circuit-switching techniques, connections
are set when needed and released when the data exchange is finished. Experiments
results are reported to show the viability of this approach, but it does not scale well as
the number of required links increases exponentially with the number of processing
elements [94]. Reconfigurable networks on chip approaches that allow runtime adap-

tation of the topology to the application requirements are proposed in [38, 139, 248]. These approaches are built on top of an FPGA-like hardware and are, however, limited to reconfigurable devices.

A NoC architecture, called ReNoC (reconfigurable NoC), is proposed in [285] to enable the network topology to be configured by the application running on the SoC architecture. More precisely, the topology can be customized for the application that is currently running on the chip by adding long and direct links between IP cores. The NoC architecture is viewed by the application as a logical topology built on top of the real physical architecture. To create these application-specific topologies, ReNoC combines packet-switching and circuit-switching in the same topology. This makes it possible to create application-specific topologies in a general NoC-based SoC platform. The evaluation studies showed that the power consumption was decreased by 56% when configuring an application specific topology compared with the static 2D mesh topology. In [131], authors evaluated a bypassing technique that allows a selected traffic to avoid the full routing functionality of selected nodes in a NoC. The results showed that this technique dramatically reduces the overall energy consumption.

1.2.2   Bandwidth allocation

Approaches dealing with topology customization assumed that all links in the NoC are identical, i.e., with uniform capacities. In NoC, if links are given the same capacity/bandwidth, some of them might be much larger than needed, consuming unnecessary power and area. Therefore, capacity or bandwidth allocation approaches that assign enough capacity to links based on the data flows have been proposed in order to decrease the latency and minimize the energy consumption. For example, an approach based on an analytical delay model for virtual channel wormhole networks with no-uniform links is proposed in [119]. This approach assigns links capacities to satisfy packet delay requirements for each flow. However, this approach considers that communication characteristics are known at design time. Approaches for dynamic link capacity allocation are required to deal with dynamic environments where communication or system characteristics are strongly changing at runtime. By increasing/decreasing links capacity, the latency can decrease.

Few approaches have been proposed to deal with dynamic link capacity allocation. For example, an approach, called 2X-Links, is proposed in [94] to allow links to change (at runtime) its supported bandwidth. Links capacity is not design-time parameterized but can be adapted at runtime depending on the current traffic. More precisely, a runtime observability infrastructure monitor the traffic and decides on when and how a certain switch should be configured. This approach was evaluated using a real-time multi-media application and the E3S application benchmark suits and reported results show that 2X-Links approach provides a higher throughput and high fault tolerance.

An adaptive approach that adjusts links bandwidth to changing network conditions is proposed in [62]. In this approach, the bandwidth of the link in one direction can be increased at the expense of the other direction. More precisely, a bandwidth arbiter governs the direction of a bidirectional link based on pressure from each node, a value reflecting how much bandwidth a node requires to send flits to the other node. A detailed simulation using a deterministic routing method on a 2D mesh and under uniform and bursty traffic was conducted, and reported results showed that the performance gains are significant.

1.2.3   Buffer allocation

Studies in NoC have shown that the buffer size significantly affects system performance, silicon area, and power consumption and increases the complexity of the system design [17,214,233]. Several approaches have been proposed to allocate buffers resources using simulation and/or analytical modeling [138,318]. However, in these approaches, optimal buffers sizes are pre-determined at design time based on a detailed analysis of application-specific traffic patterns.

Approaches to dynamically allocate buffer sizes at runtime are required for a general purpose SoCs to maximize the performance regardless of the traffic of the application. To the best of our knowledge, no research has been conducted to dynamically change the size of the buffer but some techniques have been proposed to reduce or eliminate the use of the buffers. For example, a technique called Blind Packet Switching (BPS) has been proposed in [109]. In this technique, buffers of the switch ports are replaced by simple latches and every latch can store a flit, and therefore, buffer control logic has been removed. Furthermore, latches do not store the flits while packets are blocked because they work as repeaters by storing a flit during a clock cycle. The results showed that this technique improves the performance while reducing the area and the power consumption. Another approach for routing without using buffers is proposed in [214]. The reported results show that routing without buffers significantly reduces the energy consumption, while providing similar performance to existing buffered routing algorithms at low network utilization. However, this bufferless technique delivers reasonable performance only when the injection rate is low [205]. A single-cycle minimally adaptive routing and bufferless router, called SCARAB, for on-chip interconnection networks was proposed in [122]. Reported results showed that SCARAB lowers the average packet latency by 17.6% and the energy consumption by 18.3% over existing bufferless on-chip network designs under low and moderate traffic loads.

In [159], a multi-purposing repeater logic, called IDEAL, on internodes links as storage elements is proposed in order to reduce buffering requirements. More precisely, this approach employs circuit and architectural techniques at inter-router links and router buffers, respectively. At links, repeaters are enhanced to adaptively work as buffers (i.e., serial FIFO buffers) during the network congestion. At the router buffer, architectural techniques, such as proposed in [229], are deployed for dynamic buffer allocation to prevent performance degradation. Obtained results showed that this approach achieves nearly 40% reduction in buffer power alone, more than 30% savings in the overall network power, and 35% savings in the total area.

A router architecture is proposed in [205] with a technique to dynamically tune the frequency of routers in response to network load. This technique uses the buffer utilization per port in a router as congestion metric to decide whether this port of the router is likely to get congested in the next few cycles. If so, it signals the upstream router to throttle. Using this technique, the congested router is unable to arbitrate and push out its flits fast enough compared with the rate of flit injection into its buffers. An adaptive buffer allocation scheme is proposed in [93]. The results showed that this technique increases the buffer utilization and decreases the overall buffer usage on an average of 42% in the considered case study compared with a fixed buffer assignment strategy.

1.3 Communication level

At the communication level, three main issues are defined in designing ANoC: switching, network flow control, and data routing techniques. For example, lower latency can be obtained by the implementation of an adaptive routing that avoids faulty or overloaded switches or by using an adaptive switching technique or by using a sophisticated flow control technique. Approaches proposed in the literature that address one or more of these issues are presented in the rest of this section.

1.3.1 Switching modes

Switching mode, used to specify how data and control are related, is one factor in determining the performance [275]. Two switching modes are distinguished: packet-switching mode and circuit-switching mode [258]. In circuit-switching mode, a set up connection is required to allocate a path over which all subsequent data of the connection is transported. In packet-switching mode, there is no need for a setup phase to allocate a path. Switches inspect the headers of incoming packets to switch them to appropriate output ports. However, switches have to buffer every packet to avoid deadlocks, which lead to more complex switches and increase latency. Unlike packet-switching, circuit-switching has low complexity switching nodes because their main only function is to connect an incoming link to an outgoing link. Furthermore, since the circuit setup can either succeed or fail but it cannot stall somewhere, deadlock avoidance is easily achieved [321]. Latency is dependent only on the distance but not on the network traffic.

In [321], a SoCBUS NoC is proposed and uses a hybrid packet-circuit switching method called packet connected circuit (PCC). This approach uses a small routing packet that traverses the on-chip interconnect and sets up a circuit switched route for bulk payload data to follow. In SoCBUS, when a flit is blocked, it is not stored in the switch but is dropped with the packet-switching flow control. This technique eliminates the need of buffers and thus the latency penalty. A hybrid circuit-switching technique is proposed in [144] by eliminating the circuit setup time overhead and intermingling packet-switched flits with circuit-switched flits. Authors evaluated traditional circuit-switching and compare it to a hybrid combination of circuit-switching and packet-switching. Reported results showed that this hybrid circuit-switching technique delivered the latency benefits of circuit-switching, while sustaining the throughput benefits of packet-switching with low area and power overhead. A re-configurable circuit-switched router is presented in [322]. Three applications are evaluated and the results showed that the circuit-switched NoC satisfies their requirements. Furthermore, the circuit-switched router has lower power consumption, a smaller chip area and a maximum throughput per direction compared with a packet-switched.

An approach was proposed in [5] to allow NoC to dynamically configure itself with respect to the switching modes with the changing communication requirements of the system. The main objective of this approach is to provide a low latency, a low power, and a high data throughput. This approach is decentralized because the decision to switch between packet and circuit-switching scheme is made at the router level. Therefore, the system continues performing its functions in the presence of a node failure. Simulation results illustrated that this approach uses resources efficiently compared with a traditional approach wherein all decisions are made at design time. Similarly, an approach is proposed in [206] by integrating a packet-switched and a circuit-switched network into a single architecture. Simulations results showed a promising improvement in power and performance compared with only a packet-switched NoC.

1.3.2   Routing schemes

In ANoCs, effective routing approaches can make best use of links' bandwidth and balance traffic by avoiding hot spots. Routing techniques can be classified into two main categories [258]: deterministic techniques and adaptive techniques. In deterministic routing techniques, the path between a source node and a destination node is completely determined in advance. These routing techniques are usually used because they require implementing a simple logic and therefore provide low routing latency when the network is not congested and minimize the silicon area inside the chip [46]. However, as the network becomes congested with heavy traffic, these techniques can suffer from throughput degradation and latency increases. Unlike deterministic techniques, adaptive techniques can dynamically respond to network congestion by using alternative routing paths to avoid congested links and therefore providing high performance (low latency, low packet drop, and high throughput). However, they require the implementation of a complex logic at the switch level.

Recently, hybrid techniques that combine the advantages of deterministic and adaptive techniques have been proposed. For example, an approach to switch between a deterministic routing and an adaptive routing based on the network's congestion conditions is proposed in [135]. This approach is based on odd-even routing algorithm [61] and combines the advantages of these two techniques by proposing a technique, called DyAD, from Dynamically switching between Adaptive and Deterministic modes. In this approach, a switch continuously monitors its local network load and makes decisions based on this information. In normal situations (i.e., no congestion), switches route packets using the deterministic technique and, therefore, provide low routing latency. A switch backs to an adaptive technique when congested links are detected and thus avoids these links by exploiting other paths. Simulations are conducted to compare DyAD with purely deterministic and adaptive routing schemes under different traffic patterns, and the results showed that DyAD achieves better performance only with negligible implementation overhead. Switching between deterministic and adaptive routing can lead to a higher throughput and a lower latency, which are highly desirable for NoC applications.

In [74], an adaptive approach, inspired by Ant colony [53], is proposed to route the packet traffic in order to minimize hot spots in the network. In this approach, switches select the shortest path with the least traffic for sending the packet forward. Simulations are conducted to compare the routing algorithm with other ones, such as DyAD. The reported results showed a lower average delay and less hot spots, i.e., nodes with high traffic.

A routing approach, called dynamic XY (DyXY) routing, is proposed in [184] to avoid packets to be forwarded to congested switches. By monitoring the congestion status in the proximity, routing decisions made efficiently limiting a packet to traverse the network only following one of the shortest paths between a source node and a destination node. Analytical models based on queuing theory and simulations are conducted to compare DyXY with static routing and odd-even routing algorithm for a 2D Mesh architecture. Reported results showed that DyXY outperforms these algorithms by achieving better performance in terms of latency.

In [213], a routing algorithm, called IIIModes, is proposed to allow routers to switch among three routing modes: deterministic, minimal adaptive, and non-minimal adaptive routing, based on the network's congestion conditions. A *nxn* network of interconnected tiles with a mesh topology was considered, and the simulations results showed that the average delay of IIIModes performs better under different traffic loads/patterns with low hardware overhead in comparison to other popular routing schemes (e.g., DyAD, XY).

The routing technique proposed in [258] acquires information from neighbor switches to avoid network congestion and uses the buffer levels of the downstream switches to perform the output selection. A low-latency router architecture, which utilizes an adaptive routing algorithm, is proposed in [156]. The architecture was evaluated using various traffic patterns, and results showed that this architecture reduces the overall network latency and the energy consumption while it provides, for each switch, real-time updates congestion information about neighboring switches. Similarly, a Regional Congestion Awareness (RCA) technique is proposed in [113]. In this technique, a light-weight monitoring network aggregates and transmits metrics of congestion to other switches to be aware of network hot spots.

1.3.3 Flow control schemes

Network flow control techniques address the limited buffer space in switches. In circuit-switching, there is no need of flow control techniques since the circuit is allocated at set-up phase. These techniques are required only when packet-switching mode is used because packets must be buffered before sending them to other switches or IP cores. Three categories of flow control techniques are distinguished: store-and-forward, virtual cut-through, and wormhole techniques. In NoCs, the most adopted flow control scheme is the wormhole, and it is more suitable compared with store-and-forward and virtual cut-through switching [36].

Recent studies have shown that the wormhole technique requires least buffering space, which decreases latency when there is no blocking problem. However, when the first flit of a packet is blocked, all flits of that packet are also blocked in intermediate switches [229]. It has been demonstrated that Virtual Channel (VC) flow control [71] can deal with the blocking issue in wormhole switches by assigning multiple virtual channels, each with its own associated buffer queue, to the same physical channel. In other words, VC flow control is based on organizing buffers associated with an input port as parallel queues (virtual channels) that can be allocated to arriving flits. Flits are forwarded in the network over the virtual channels; when a flit is blocked in one channel, the other virtual channels can still use the physical channel. As demonstrated in [241], VC flow control increases throughput by up to 40% over wormhole routers. However, design time allocation of VCs structure lacks of flexibility on various traffic conditions and corresponds to low buffer utilization. Therefore, dynamic virtual-channel flow control methods have been proposed to tackle this issue.

In [229], a dynamic Virtual Channel Regulator (ViChaR) is proposed to dynamically allocate virtual channels and buffer resources according to network traffic conditions. In this technique, the number of VCs and the buffer size per VC are dynamically adjusted based on the traffic load. The reported results showed that ViChaR achieved similar performance with half the buffer size of a generic router. While ViChaR can achieve good performance, VC arbitration and its logic are more complicated, and increasing the number of VCs arbitrarily can increase latency. This issue was motivated the work, which have been done in [158] by adopting a dynamic VC table-based approach with a fixed number of VCs. The reported results showed a 30% reduction in overall network power and a 41% reduction in area when half of the input buffers in the router are removed.

A dynamically allocated VC approach with congestion awareness is proposed in [166]. This approach extends deep VCs at low traffic rate in order to reduce packet latency, while it increases VC number and avoids congestion situations to improve throughput in high rate. Simulations are conducted and results are reported and showed that at different injection rates or traffic patterns this dynamic allocation approach provides on average 8.3% throughput increases and 19.6% latency decreases.

In [255], a dynamic power management technique, called dynamic virtual channels allocation (DVCA), for optimizing the use of virtual channels is proposed. This technique optimizes the number of active VCs for the router input channel based on the traffic condition and past link utilization. More precisely, based on the predicted traffic, the number of active virtual channels is adjusted, i.e., increased, decreased, or remain the same. Simulations results showed that up to 35% reduction in the buffer power consumption and up to 20% savings in the overall router power consumption are achieved.

A flow control technique using the storage, which is already present in pipelined channels in place of explicit input virtual channel buffers, is proposed in [202]. The results showed that using elastic buffer improves the performance compared with virtual channel usages. In [164], a control flow mechanism, called express virtual channels (EVCs), was proposed to allow packets virtually bypassing intermediate routers along their path. Reported results showed up to 84% reduction in packet latency and up to 23% improvement in the throughput while reducing the average router energy consumption by up to 38% over an existing state-of-the-art packet-switched design. According to these studies, virtual-channel flow control methods are more suitable in NoCs because they require minimum buffer space while offering high throughput and low latency and energy consumption.

1.4 Application level

Mapping and scheduling the application tasks into NoC platforms while optimizing cost metrics and performance metrics constitutes the main issue at this level [233]. Mapping approaches define which processing element or IP core executes certain application tasks, but tasks scheduling is another issue that refers to the order in which tasks are executed. Approaches for applications mapping and scheduling can be classified into two main categories: design-time approaches and runtime approaches. Design-time application mapping and scheduling was addressed by many researchers in the past and several approaches have been proposed, such as in [134, 136, 177, 219, 324]. These approaches are not appropriate for dynamic workloads because in dynamic NoC, it is unknown which applications run simultaneously and in which circumstances with regard to available resources and end-user behavior [54, 95, 278]. Runtime mapping/scheduling approaches are required for applications with unpredictable workload like, for example, user-induced multimedia processing, and subsequently unpredictable changes in active tasks and communication requirements [337].

Run-time approaches proposed in the literature can be classified into two main classes, centralized and decentralized. Centralized approaches used centralized controller or manager for monitoring traffic and conducting the mapping process. For example, an approach for run-time task assignment on heterogeneous processors is proposed in [278]. The obtained results showed that this approach provides a near optimal solution in a reasonable short computation time. In [54], the performance of several run-time mapping heuristics with dynamic workloads was investigated. Based on communication requests and the load in the NoC links, tasks are mapped on the fly. Reported results showed a reduction in execution time and network congestion. However, these centralized approaches may not be scalable in the context of large NoCs with hundreds or even thousands of cores. They present a single point of failure and larger volume of monitoring traffic [95].

Decentralized approaches, however, use multiple entities to perform the traffic monitoring and mapping. For example, a decentralized approach is proposed in [95] using virtual clusters that are constructed at run-time. The clusters are connected subsets of NoC tiles and have a variable size that may be adjusted at runtime. Each cluster has one component that is responsible for (re-)mapping. In [337], a decentralized heuristic algorithm is proposed to allow each IP core/processing element to migrate individual tasks to neighboring ones based on the local workload, task sizes, and communication requirements of the tasks to be migrated.

A hierarchical agent-monitored network-on-chips was proposed in [257] to provide diagnostic services to the system against failures or errors. Based on the performance provided, agents decide on an optimal improvement mechanism, either replacing the poor functioning processing element or adding more processing elements, or speeding up the circuits. In this approach, agents are imitating the adaptive behaviors observed in biological systems by autonomously adjusting the system performance at their own level. For example, higher level agents are supervising over lower level agents so that the whole system can achieve the optimal overall performance. The objective of this work is toward developing an autonomic NoC with self-aware properties into component level and strengthen self-design and fault-tolerance aspects.

A service-oriented organic real-time middleware, called CARISMA, was proposed in [227]. This middleware forms a homogeneous layer on top of the interconnected nodes and allows to run system and application services. In their approach, tasks are allocated autonomously to the most suitable node, and during run-time, continuous reallocation and reconfiguration shall maintain a near-optimal system state. Furthermore, tasks are able to move freely between appropriate cores with the objective to improve the overall performance. As pointed out in [233], the task and communication scheduling/binding and the IP mapping should be performed in parallel, and significant work is required to develop run-time and efficient performance and energy-aware approaches.

1.5 BNoC Architecture

The NoC infrastructure is the combination of various elements (e.g., switch, links) and protocols (e.g., routing, switching) that determine the communication architecture and modes. In open environments, with the rising need for on-demand services, a high degree of self-management and automation is required for ANoC infrastructures. NoC must be enhanced by adaptive capabilities, such as self-configuration, self-optimization, and self-healing. For example, NoC elements must be able to change the traffic route at runtime in order to efficiently avoid faulty areas or hot spots. Therefore, ANoC infrastructures that provide these self-* features in all levels, as described already, are required. These self-* features have been mainly studied to develop large and adaptive distributed systems [10, 208]. Furthermore, this is in part the aim of autonomic computing, which was an IBM initiative to deal with IT servers complexity [152] by developing systems that mange themselves. In other words, autonomic computing focuses on creating computer systems that manage themselves according to an administrator's goals. The major characteristics of autonomic computing are self-configuration, self-healing, self-optimization, and self-protection.

In the rest of this section, an architecture inspired by the immune system for ANoC is introduced. First, an overview of immune system principles is presented by describing the structure of the immune system from a biological perspective. The immune system has a useful set of organizing principles, such as self-configuration, self-optimization, and self-healing principles, that can guide the design of ANoC. The BNoC is then presented together with some preliminary results.

1.5.1   Immune system principles

The immune system defends the body against harmful diseases and infections [279]. Once pathogens enter the body, they will be handled by two subsystems, the innate and the adaptive immune system, which work together to achieve this task. The innate and adaptive immune systems are produced primarily by leukocyte cells. Among the several different types of leukocytes, there are phagocyte and lymphocyte cells. The phagocyte cells are the first line or level of defense for innate immune system. They engulf the pathogen and present it to the adaptive immune system. More precisely, the innate immune system hosts defense in the early stages of infection through nonspecific recognition of a pathogen and inhibits the adaptive immune response.

The adaptive immune system primarily consists of lymphocytes that circulate through the body in the blood and lymph networks. There are two categories of lymphocytes, the T-cells and B-cells. The role of T-cells is to potentiate the immune response by the secretion of specialized factors that activate other cells to fight off infection. The major function of B-cells is the production of antibodies in response to foreign antigen. The main characteristic of the adaptive immunity is the specific recognition of pathogens leading to the generation of pathogens specific response.

After producing the immune response, a feedback is introduced to the tissue to induce healing and to the immune system itself to modify its structure and its future behavior [69]. Therefore, the immune response is formulated a priori by the immune system in response to the cumulative experience of the immune system in dealing with the body (the self) and with the world (the foreign).

Another mechanism proposed by Jerne in [145], called immune network theory, in which B-cells are not isolated but form an idiotypic network for antigen recognition. These cells both stimulate and suppress each other in certain ways that lead to the stabilization of the network [76]. Two B-cells are connected if the affinity they share exceed a certain threshold, and the strength of the connection is directly proportional to the affinity they share. In this model, T-cells help is never a limiting factor for B-cell proliferation or production of antibodies.

Another model was proposed in [306], called second generation immune system, in which the activation of B-cells is explicitly dependent on their cooperation with activated T-cells. The antibodies produced by the B-cells mediate the idiotypic interactions between them and controls their induction. In fact, circulating antibodies are the only inhibitory influence on T-cells activation and growth. A bounded dynamics of the T-cell activity can be achieved if and only if their receptors are integrated into the idiotypic network.

Immunological principles and functionalities from computational viewpoint have been applied in many domains [76, 101, 121]. For example, in [101], an interesting model inspired by the natural immune system to monitor resources and services in distributed networks was proposed. The author has suggested that an intelligent network with self-organizing and emergence capabilities can react like the natural immune system against pathogens that have entered the body. The mapping is the following, pathogens (i.e., virus) correspond to user requests and immune responses correspond to request resolutions. More precisely, each user request is considered as an attack launched against the global network. The intelligent middleware detects the infection (i.e., user request) and delivers a response to eliminate it (i.e., satisfy the user request).

The biological immune system has several features and organizing principles that can be exploited to develop autonomic systems [121]. The immune system can be seen as a massively parallel architecture with a diverse set of cells, which are distributed throughout the body but can communicate using chemical signals. There is no central control, the multitude of independent cells work together resulting in the emergent behavior of the immune system. Furthermore, the immune system is multilevel and each level operates in concert with all the other components to provide defense-in-depth. The immune system evolves to adapt and improve the overall system performance.

1.5.2   BNoC principles

As described already, a system-on-chip can be viewed as a micronetwork of components in which the OCI is the abstraction of the communication among PEs and cores. The OCI must satisfy quality-of-service requirements, such as reliability, performance, and energy/area bounds. By analogy to the biological immune system (BIS), BNoC can be seen as a logical system hierarchy of levels, application level, communication level, and architecture level. The OCI corresponds to the lymph and blood networks, immune cells correspond to software components or agents, body cells correspond to NoC components (i.e., PEs or cores). As in BIS, BNoC operations imply a suite of functions performed by a federation of interacting components or agents. Structural features of the application to be maintained are represented as antigens to be cleared, while its behavioral features are represented as signals. The immune response against the antigen corresponds to the techniques selected to adapt to the behavioral/structure of the application [18]. The mapping between these functions and organizing principles can be seen in Table 1.1.

At each level, monitoring and supervising mechanisms are required to inhibit the corresponding self-* feature. Using the micronetwork stack model, the self-* capabilities can be defined as follows (see Figure 1.2). This allows higher level components to allocate and supervise tasks of lower level agents. These lower level agents

TABLE 1.1

The mapping between BIS and BNoC

BIS	BNoC
Tissue	NoC
Body cells	PEs/cores
Lymph/blood streams	OCI
Immune cells	Components/agents
Antigens	Tasks/system status
Immune response	Adapt to changes

BNOC	Self-* Capabilities	Immune System
Application level - Mapping - Scheduling	Self-configuration - Adapt to changes	Innate level - Engulf pathogens - Activate T-Cells
Communication level - Routing - Switching - Flow control	Self-healing - Act to the disruption	Adaptive level (T-Cells) - Learn Ag structure - Activate B-Cells
Architecture level - Buffer - Bandwidth - Link	Self-optimizing - Tune resources	Adaptive level (B-Cells) - Learn Ag behaviour - Produce antibodies

FIGURE 1.2

The micronetwork stack functions with self-* capabilities.

monitor the local PEs and adapt to the task's behavior. They also report the changes made at this level to the higher level agents.

Alike immune system, three types of agents are considered, P-agents, T-agents, and B-agents. At application level, innate immune components must be *self-configuring*. Similar to phagocyte cells that engulf the pathogen into antigens and present them to adaptive immune system (T-Cells), the P-agents analyze the application requirements, decompose them into tasks, and map them by presenting them to corresponding T-agents. Furthermore, the P-agents should dynamically adapt to changes in the environment, such as dramatic changes in the system characteristics or the user behavior by deploying new tasks or the removal of existing ones. This dynamic adaptation helps ensuring continuous operation of the application under unknown circumstances.

At the communication level, similar to T-cells, located in the thymus, to coordinate and regulate B-cells activation, T-agents are regional monitor to allocate and schedule the work of B-agents by selecting the required routing/control flow technique and configuring the network within the region. T-agents have the *self-healing* capability, which allows them to discover, diagnose, and react to disruptions (e.g., deadlocks). This can occur, for example, when a group of flits are blocked and cannot make progress because they are waiting for one another to release buffers or channels. In these situations, T-agents monitor and transmit metrics of congestion throughout to the other T-agents to be aware of network hot spots and therefore avoid them by selecting another route. For example, based on the changing communication requirements of the system, T-agents can dynamically switch between the packet-switching scheme and circuit-switching scheme.

At the architecture level, similar to B-cells that are local monitors and change their behavior to adapt to the antigen behavior, B-agents should implement various techniques for resources tuning, such as buffer allocation and bandwidth adaptation. In other words, B-agents have the *Self-optimizing* capability that allows them to automatically monitor and tune resources with the objective is to increase the performance while lowering area and energy consumption. For example, based on the predicted traffic, the number of the active virtual channels can be adjusted to reduce in the buffer power consumption.

1.5.3   Simulation results

In this section, some preliminary results are presented by analyzing the delay, the throughput, the communication load, the loss rate, and the energy consumption experienced by data flows sent by the PEs. In the simulation, the application is represented as communicating parallel processes already mapped into PEs. Each process is linked with a traffic generator that injects flits according to the CBR (Constant Bit Rte) model at a deterministic rate, which is varied from light traffic (25 Mbps) to heavy traffic (100 Mbps). The links' bandwidth is fixed to 200 Mbps, the buffer size is fixed to 16 flits, the flit size is 8 bytes, and the link delay is 320 μs [291].

The behavior changes of the application are modeled by increasing or decreasing injection rate. The *self-healing* mechanism is first developed to allow for some sources/sinks, represented by B-agents, to establish affinity links in order to adapt to the application behavior. By monitoring and controlling data flow on a feedback basis of B-agents, which monitor switches queues, if a buffer is about to overflow (greater than a fixed threshold), then T-agents inform the B-agent source and B-agent sink of this traffic to establish an affinity link to avoid hot spots, similar to the affinity link or relationship in immune system. Affinity link can be imitated by establishing a direct link between the source and destination nodes or by a circuit such as in circuit-switching scheme.

To illustrate the benefit of affinity links establishment, we consider the WK onchip interconnect [291] as a case study (see Figure 1.3). All PEs are traffic sources that transmit flits. However, sinks (destination PEs) are selected according to the following communication locality principle in which 25% of the traffic takes place

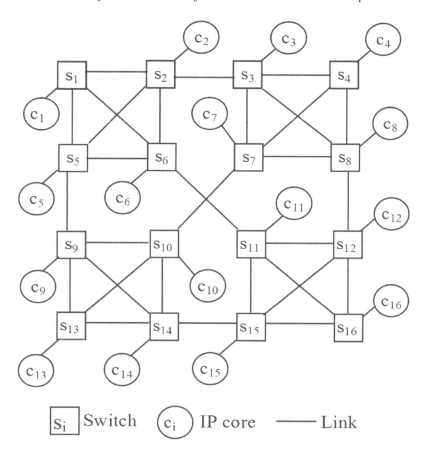

FIGURE 1.3
WK-recursive on-chip interconnect.

between neighboring cores and 75% of the traffic is uniformly distributed among the rest. In this simulation, all PEs are selected to be sources and sinks in order to simulate high traffic network load.

Figure 1.4(a) shows the average end to end with and without creating affinity links. In this figure, the average delay increases linearly as the injection rate is above 50 Mbps. When increasing the injection rate, the network becomes more congested with heavy traffic (see Figure 1.4(b)). Buffers become full causing more flits to wait and so the average latency increases. In this figure, without creating affinity links (S0), the latency is high because of packet-switching overhead inside switches (e.g., buffering, arbitration, routing). However, establishing affinity links (S1) between nodes decreases the latency because this allows bypassing intermediate switches and therefore decreasing the amount of traffic in these switches. Results show a signifi-

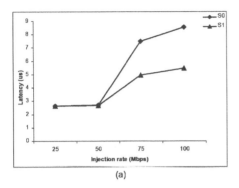 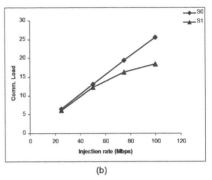

(a)   (b)

FIGURE 1.4

(a) Average latency and (b) communication load (%) vs. injection rate.

cant improvement (up to 35% in heavy workload) when appropriate affinity links are created.

The communication load is another metric we used to show the benefit for creating affinity links between nodes. It is a relative value of arrival rate versus departure rate on all links. Figure 1.4(b) shows that when increasing the injection rate the communication load increases. The sensitivity to traffic increases is less when adding affinity links (up to 21% at injection rate 100 Mbps). When adding affinity links (S1) between nodes, the traffic is routed through these links and, therefore, the communication load is decreased compared with static topology (S0).

Figure 1.5 illustrates the variation of loss rate (i.e., flits not forwarded due to lack of buffer space) under different injection rates. Without creating affinity links (S0), as the injection rate is above 50 Mbps, the loss rate increases because the network becomes more congested with heavy traffic so queues become full causing more flits to drop. When establishing affinity links (S1), no flits loss is detected as the injection rate increases.

Aggregated throughput is a metric that represents how many bits arrived at destination nodes per second. Figure 1.6(a) shows the variation of aggregated throughput under different injection rates. This figure shows that when the injection rate is above 50Mbps, the improvement is up to 9% on average because of lower loss rate as depicted in Figure 1.5. When the loss rate decreases, more flits arrived to destination nodes.

The results depicted in Figure 1.6(b) compares the energy consumption in function of the injection rate. This figure shows that the average energy consumption increases linearly when the injection rate increases. This increase can be explained by the huge number of flits generated as the injection rate increases. Without creating affinity links (S0), we can see highest energy consumption in heavy traffic (injection rate is above 50 Mbps) because flits have to be buffered and traverse more hops. Notice that similar figures were produced using different buffer sizes and results related to these performance metrics showed the benefit of creating affinity links.

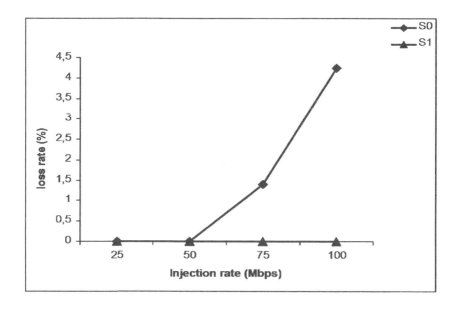

FIGURE 1.5
Flits loss rate vs. injection rate.

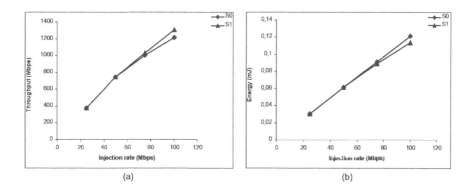

FIGURE 1.6
(a) Aggregated throughput and (b) energy consumption vs. injection rate.

1.6 Conclusions

This chapter gave an overview of state-of-the-art approaches for ANoC. Unlike design-time approaches in which all parameters/protocols are optimized/selected at design-time targeting a specific application, in run-time approaches, the adaptation process is running continuously to evolve the system. Based on this state-of-the-art review, the self-* capabilities were defined in the context of ANoC. These capabilities can provide the core scheme to develop ANoC at application, communication, and architecture level. An approach inspired by biological immune system toward developing ANoC with these self-* capabilities is introduced. One aspect of the self-healing capability is partially developed, and some preliminary results are reported and showed that the adaptation is energy and latency efficient. Despite these promising results, the proposed approach is still in its infancy and can be improved by developing and integrating all self*- capabilities into one NoC platform.

2

Bio-Inspired NoC Architecture Optimization

A. A. Morgan

Department of Electrical and Computer Engineering, University of Victoria, BC, Canada

H. Elmiligi

Department of Electrical and Computer Engineering, University of Victoria, BC, Canada

M. W. El-Kharashi

Department of Computer and Systems Engineering, Ain Shams University, Cairo, Egypt

F. Gebali

Department of Electrical and Computer Engineering, University of Victoria, BC, Canada

CONTENTS

One of the challenging problems in Network-on-Chip (NoC) design is optimizing the architectural structure of the on-chip network in order to maximize the network performance with the minimum possible cost. Recently, NoC implementations moved from using standard communication architectures to semi- or fully-custom ones. In this chapter, we present a multi-objective optimization of NoC architectures using bio-inspired techniques. According to the application description, the NoC architecture generation problem is formulated as a multi-objective optimization one. Thereafter, a genetic algorithm (GA)-based technique is presented to solve this optimization problem. Our technique is evaluated by applying it to different NoC benchmark applications, as case studies. Results show that the architectures generated by our technique outperform other standard and custom architectures with respect to power, area, delay, reliability, and the combination of the four metrics. Moreover, the low running time complexity of our technique makes it more attractive for Autonomous NoC (ANoC) compared with other architecture optimization techniques.

2.1   Introduction

Networks-on-Chip (NoC) was proposed at the beginning of this century to handle the communication problem between large number of computational resources in modern applications [29]. NoCs first employed predefined 2D or 3D general-purpose standard architectures, used in computer networks and multi processor systems, to realize NoC-based designs [237, 239]. However, as modern application requirements differ significantly, the use of standard architectures might not guarantee a specific level of performance. Therefore, *Application-specific Networks-on-chip (ASNoC)* was presented to customize NoC architectures according to application requirements and design constraints [27, 234, 284]. Moreover, in recent years, *Autonomous Networks-on-Chip (ANoC)* started to become a hot research area [294]. In the literature, ANoC is used to represent two types of on-chip networks. The first is the *fault-tolerant network*, which is capable of operating properly even in the event of components failure [304]. The second is the *adaptable self-optimized network* that could adjust itself dynamically to optimize certain performance and cost metrics [329]. In this chapter, we target these self-optimized networks by presenting a new methodology to optimize the underlying on-chip network architecture during runtime.

The design of ANoCs requires implementing three *basic modules*. These are the monitoring, the architecture optimization, and the dynamic reconfiguration modules, respectively [115]. The first module monitors the changes in both external user-supplied and internal on-chip running parameters. Examples of these parameters are the application requirements, the design constraints, the inter-cores traffic, and the on-chip noise. Real-time monitoring has been previously addressed in several research works [66, 115, 329]. Secondly, the architecture optimization module generates the optimum network architecture based on the inputs from the monitoring module. Previous works in NoC architecture generation consider only one objective for optimization [172, 284]. Moreover, the running time of these techniques prevents them from being used with ANoCs. Accordingly, in this chapter, we address these problems by presenting a biologically inspired multi-objective technique to quickly generate the best NoC architecture for certain given design parameters. Finally, the reconfiguration, or the action, module reconfigures the network to the new architecture generated from the optimization module. Similar to the real-time monitoring, NoC reconfiguration is addressed by many previous research works. For example, dynamic reconfiguration for Field Programmable Gate Arrays (FPGAs) is discussed in [124, 247]. Moreover, platform-based design for reconfigurability is presented in [207].

For any NoC-based system, the on-chip network architecture significantly impacts its cost and performance [172, 237]. On the cost side, it is required to minimize the NoC area and power consumption. On the performance side, it is required to minimize the delay and maximize the throughput and the reliability. However, performance requirements and cost constraints are usually conflicting. For example, increasing the NoC reliability often requires using more resources, which increases the NoC area and power consumption. Accordingly, modern NoC research proposes different methodologies for customizing NoC architectures according to the application requirements. However, works done so far customize the NoC architecture for a single metric, which is the most important for the application, assuming other metrics as constraints [85, 172, 234, 284]. Unfortunately, optimizing for a single metric usually affects the performance with respect to other metrics. Therefore, one challenging problem is to optimize on-chip networks for more than one metric. That is to decide, at both system and circuit levels, which NoC architecture results in the maximum possible performance and the minimum possible cost simultaneously.

In this chapter, we present using bio-inspired techniques to solve the above mentioned multi-objective optimization problem. More precisely, we employ GA-based multi-objective optimization to achieve the best custom architecture for the application of interest. Moreover, our technique trades off many conflicting NoC design variables. The most important of these variables are the number of ports per router, the traffic per port, the traffic traces routing, the average number of hops, and the probability of the traffic being affected by on-chip noise sources.

The remaining of this chapter is organized as follows. Related work is discussed in Section 2.2. Section 2.3 surveys bio-inspired optimization techniques. Section 2.4 gives some background from graph theory on how to represent different NoC applications. The performance and cost models used in this chapter and the formulation of the multi-objective optimization problem is presented in Section 2.5. Section 2.6 presents our GA-based solution for the custom architecture generation problem. Section 2.7 presents some experimental results for different NoC benchmarks to validate our work. Finally, we draw conclusions and give ideas for future work in Section 2.8.

2.2 Related work

The work done for NoC *architecture generation* so far could be classified into three main research directions:

1. Employment of known *standard architectures* within general-purpose NoCs,

2. Generation of *semi-custom architectures* to enhance some performance metrics, and

3. Generation of *fully-custom architectures* that are optimized for a specific metric according to the on-hands application needs.

2.2.1   Employment of known standard architectures

Early research works in NoC employed predefined 2D architectures, used in computer networks and multi processor systems, to realize NoC-based designs [30, 36]. At a more advanced stage, researchers used 3D architectures to generate better-performance on-chip networks [239]. Analysis and evaluation of different standard architectures with respect to throughput, latency, energy consumption, and area have been done by many researchers. For example, Pande et al., in [237], discussed and compared six different architectures (SPIN, mesh, torus, folded torus, octagon, and BFT) with respect to all previous metrics. The same was done in [41] for ring, spidergon, and mesh architectures with respect to throughput and latency only. Moreover, different algorithms and heuristics were proposed to map application cores onto these standard architectures. The most famous of these algorithms are PBB [133], GMAP [133], PMAP [161], NMAP [220], and BMAP [274]. Finally, some researchers used different bio-inspired techniques, like Evolutionary Algorithms (EAs) and Simulated Annealing (SA), to obtain the optimum mapping onto only 2D mesh architecture [8, 190, 334, 335].

The problem of choosing the right standard architecture for NoC-based systems was addressed in [221]. Murali and De Micheli presented a tool, SUNMAP, to automatically select the best architecture for a given application and map its Intellectual Property (IP)[1] cores onto that architecture. Unfortunately, SUNMAP is limited to only five standard architectures (mesh, torus, hypercube, butterfly, and clos).

2.2.2   Generation of semi-custom architectures

The research work in this area could be considered as the intermediate stage between standard general-purpose architectures and application-specific fully custom architectures. As the NoC-based design paradigm matured, researchers realized that the NoC domain is different from the computer network domain. While the latter has a large amount of uncertainty, the former is nearly deterministic. The NoC design objectives and the application traffic are usually known, or could be precisely estimated, a priori. Therefore, researchers started customizing the NoC architecture to comply with the application requirements. One research direction for this partial customization tried to slightly modify standard 2D mesh architecture to enhance certain metrics [234, 316]. Another direction tried to combine more than one standard

---

[1] IP is a parameterizable core that can be used in SoC design implementation.

architecture to generate a semi-optimized architecture with respect to certain metrics [91, 155, 209].

2.2.3 Generation of fully-custom architectures

Recently, researchers started to move from just enhancing the performance of the on-chip architecture to optimize it with respect to a specific metric [172, 283]. The resultant architecture is completely custom and outperforms other architectures with respect to the metric for which the optimization is carried out. However, the direct impact of the single-objective optimization is the deterioration that happens to other metrics. Therefore, we presented a two-objective optimization methodology in [212] to generate the best NoC architecture with respect to both area and delay. Moreover, in this chapter, we build on our previous work and present the use of bio-inspired techniques for multi-objective optimization of NoC architectures. Our technique considers four NoC metrics: power, area, delay, and reliability. Furthermore, the assumptions used in our technique aim at reducing its execution time to make it suitable for ANoC.

2.3 Bio-inspired optimization techniques

Bio-inspired *optimization* techniques are those that mimic the natural biological systems. These techniques are proposed to solve the real complex problems that could not be solved by conventional optimization techniques. *Bio-inspired techniques* could be classified into two main categories: *Evolutionary Algorithms* (EAs) and *Swarm Intelligence* (SI). The former were inspired by genetic evolution, whereas the latter tried to mimic animal behavior. The most famous EA techniques are GAs [108], Genetic Programming (GP) [160], and Evolutionary Programming (EP) [199]. The most widely used SI techniques are Particle Swarm Optimizer (PSO) [151], Ant Colony Optimizer (ACO) [40], and Group Search Optimizer (GSO) [123]. As we are using GAs in this chapter, the following paragraphs give some details about them. For more information about the GAs and the others, the reader is referred to [293, 330].

GAs are inspired by the process of natural evolution. Accordingly, any potential solution is represented in the form of a chromosome. A set of chromosomes constitutes a generation. The algorithm adopts a stochastic global search method to evolve a new generation from the current one. The first phase of the algorithm is the *selection* in which the most fitted chromosomes from the current generation are selected. Fitness is evaluated based on an objective function that models the optimization problem. Thereafter, a new generation is produced from the selected chromosomes by applying different genetic operators:

- *Elitism*: This ensures that the best individual will survive to the next generation. Accordingly, at least, the most fitted chromosome is copied without changes from the current generation to the new one.

- *Crossover*: Chromosomes that are previously selected are considered as parents and are allowed to mate. Accordingly, parts of different chromosomes are exchanged together to produce two new children, or offspring. Crossing over good chromosomes likely results in better offspring. Finally, these newly created offspring are added to the next generation.

- *Mutation*: Selected chromosomes from the current generation are slightly altered to introduce some diversity in the new generation. This diversity prevents the chromosomes from becoming so similar to one another. Consequently, the algorithm is likely protected from being trapped in local minima or maxima.

The algorithm continues going in an iterative manner to evolve good individuals from one generation to the next until the best solution is reached. Finally, a stopping criterion should be employed to stop the algorithm.

2.4   Graph theory representation of NoC applications

Graph theory concepts could be used to represent NoC applications. The application description is assumed to be known a priori in this chapter. Moreover, *traffic* characteristics are expected to be collected and analyzed by the on-chip monitoring module. Accordingly, an application could be described during any period of time by a *core graph* [210, 221]. Core graph represents the processing elements, or cores, in the application and the traffic between them. As shown in Figure 2.1, a core graph is a directed graph G(V, E), where each vertex $v_i \in$ V represents a core in the application, and the directed edge $e_{ij} \in$ E represents the communication between cores v_i and v_j. The weight of an edge expresses the average traffic in MB/s exchanged between cores v_i and v_j. Using NoC terminology, a core graph could be represented in the form of a *traffic distribution matrix* (Λ) [88]. If there are N cores in the application, the dimensions of the matrix is $N \times N$. Any element λ_{ij} in the matrix represents the weight of the edge e_{ij}, i.e., the average traffic sent from core v_i to core v_j in MB/s.

2.5   Problem formulation

In this section, we first present the cost and performance models employed in this chapter. Models are then combined together to constitute an optimization objective function. In this objective function, a weight factor is used for every metric to allow the designer to control the optimization process according to the importance of these metrics to the application in hands. This objective function represents the fitness function of our bio-inspired technique. Consequently, our GA-based optimizer aims at minimizing this objective function.

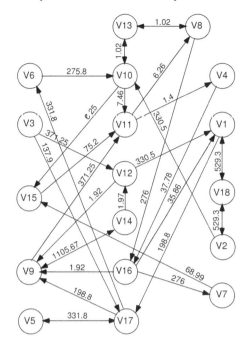

FIGURE 2.1
The core graph of the AV benchmark [85].

The running time of the optimization technique is very crucial for ANoCs. Although modifying the underlying NoC architecture is not going to be done very often, the new architecture should be generated as quickly as possible. Therefore, to ensure low execution time of our technique, we consider the following assumptions:

1. The application, with its traffic characteristics, is represented at any period of time by a core graph,

2. The generated architecture is an indirect network [30] for which each node is connected to only one router and each router has a maximum of one node connected to it, i.e., each core has its own router,

3. Each link consists of two channels for packet transmission and reception, respectively. The channel width of all the links in the network is the same and is equal to the flit size (8-bit flits are assumed in this chapter),

4. The length of all links is the same and allows for a single clock cycle data transfer, i.e., no repeaters are required. This assumption might not be achieved in the actual circuit-level floorplanning. However, it is still valid for the purpose of system-level evaluation and comparison, and

5. Deterministic shortest path routing is used.

The above assumptions aim mainly at keeping the basic underlying infrastructure of the on-chip network fixed. The advantages of preserving the same basic infrastructure are threefold. First, it reduces the reconfiguration time and the effort required for the on-chip reconfiguration module to carry out its task. Second, it enables us to use more time-efficient data structures in our optimization technique, like matrices. Third, it reduces the number of independent design variables to optimize, like the number of routers and the channel width. The latter two advantages clearly allow the architecture optimization module to achieve its goal in a short and reasonable time.

2.5.1 Power model

NoC power is consumed by its routers and links. To model the *router power*, we implement a library of routers with different number of ports and buffer sizes. Ports within the same routers have equal buffer sizes and each of them consists of two channels for packets reception and transmission, respectively. A round robin scheduler serves backlogged queues at the output one after another in a fixed order [91]. Thereafter, power simulations are done at various operating frequencies and target technologies. For example, at an operating frequency of 500MHz, Table 2.1 shows sample results of the routers power consumption with different flit injection rate, in *flit/cycle*, for 0.18μm technology. The last row represents the router leakage power only.

TABLE 2.1
Power consumption of 4-, 5-, 6-, 7-, and 8-port NoC routers for various flit injection rates at 500 MHz when implemented in 0.18 μm technology

Flit injection	Total Power (P_R) (mW)				
rate(flit/cycle)	4-port	5-port	6-port	7-port	8-port
1.000	64.104	96.885	136.044	137.379	234.287
0.400	32.019	48.440	68.041	86.709	117.173
0.200	12.793	19.380	27.229	34.706	46.901
0.100	6.410	9.705	13.635	17.372	23.481
0.050	3.211	4.862	6.832	8.705	11.762
0.020	1.293	1.963	2.747	3.505	4.726
0.002	0.135	0.203	0.285	0.38	0.487
0.000	0.008	0.013	0.018	0.025	0.032

From these results, we notice linear dependencies of both the dynamic and the leakage power on the number of ports and the flit injection rate. Accordingly, in our model, we use linear relations to calculate the power of any router with any number of ports and flit injection rate. As a result, NoC routers power is represented as

$$P_R = \sum_{i=1}^{N} P_{Ri} \qquad (2.1)$$

where

$$P_{Ri} = P_{Ri-Dynamic} + P_{Ri-Leakage} \qquad (2.2)$$

$$P_{Ri-Dynamic} = (C_{DM} \cdot M_i + C_D) \cdot \alpha_{fi} \qquad (2.3)$$

$$P_{Ri-Leakage} = (C_{LM} \cdot M_i + C_L) \cdot \alpha_{fi} \qquad (2.4)$$

where N is the number of routers in the network, which is equal to the number of cores according to our assumptions. $P_{Ri-Dynamic}$, $P_{Ri-Leakage}$, and P_{Ri} are the dynamic, leakage, and total power of router i, respectively. M_i and α_{fi} are the number of ports and the flit injection rate for router i. For any architecture, the flit injection rate could be calculated for each router by knowing the NoC operating frequency, the employed routing strategy, and the traffic distribution matrix (Λ) discussed in Section 2.4. Finally, C_{DM}, C_D, C_{LM}, and C_L are the dynamic port-dependent, the dynamic port-independent, the leakage port-dependent, and the leakage port-independent power constants, respectively. These constants are technology-dependent and could be obtained by linear curve fitting of the power simulation results of the employed routers.

Link power is calculated using the methodology presented in [91]. To send certain traffic from core v_i to core v_j, the power consumed in links depends on the amount of traffic transferred, λ_{ij}, the physical characteristics of the links, and the number of hops between the source and the destination cores. Link physical characteristic could be obtained from the corresponding technology library. Moreover, the number of hops between any two cores could be obtained from the architecture *connectivity matrix* (C). Any element c_{ij} in the connectivity matrix represents the number of hops between routers R_i and R_j. The connectivity matrix, in turn, is calculated from the *adjacency matrix* (A) of the architecture. For an application with N cores, the adjacency matrix is a binary $N \times N$ matrix, whereas $a_{ij} = 1$ only if there is a direct link connecting between routers R_i and R_j. Figure 2.2 shows an example for the adjacency and the connectivity matrices of a 6-node ring architecture. In this chapter, the connectivity matrix is calculated from the adjacency matrix using the Dijkstra shortest path algorithm [79]. Accordingly, NoC *link power* is represented as

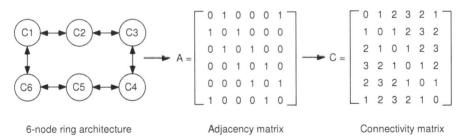

6-node ring architecture   Adjacency matrix   Connectivity matrix

FIGURE 2.2
Example of the adjacency and the connectivity matrices for a 6-node ring architecture.

$$P_L = \sum_{i=1}^{N} \sum_{j=1}^{N} \lambda_{ij} \cdot c_{ij} \cdot P_{unit\ link} \qquad (2.5)$$

where

$$P_{unit\ link} = P_{switching} + P_{short} + P_{Leakage} \qquad (2.6)$$

$$P_{switching} = \frac{1}{2} N_C \cdot V_{dd}^2 \cdot (C_L \cdot \alpha_L + C_C \cdot \alpha_C) \cdot f \qquad (2.7)$$

$$P_{short} = N_C \cdot \tau_{sc} \cdot \alpha_L \cdot V_{dd} \cdot I_{short} \cdot f \qquad (2.8)$$

$$P_{Leakage} = N_C \cdot V_{dd} \cdot (I_{bias,wire} + I_{leak}) \qquad (2.9)$$

where N is the number of cores in the application. λ_{ij} is the traffic sent from core v_i to core v_j in MB/s. Similarly, c_{ij} is the number of hops between the same two cores. $P_{unit\ link}$ is the power consumed in one unit link. $P_{switching}$, P_{short}, and $P_{Leakage}$ are the link switching, short circuit, and leakage power, respectively. The summation of $P_{switching}$ and P_{short} represents the dynamic power consumed in a link. N_C is the number of wires in a link, i.e., channel width, and V_{dd} is the supply voltage. C_L and C_C represent self and coupling capacitance of a wire and with neighboring wires, respectively. Similarly, α_L is the switching activity on a wire and α_C is the switching activity from the adjacent wires. f denotes the operating frequency and τ_{sc} is the short circuit period during which I_{short} flows between source and ground. Finally, $I_{bias,wire}$ represents the current flowing from the wire to its substrate and I_{leak} is the leakage current flowing from the source to ground regardless of the gate's state and switching activity. All the parameters for link power calculation could be obtained from the Predictive Technology Model (PTM) [250].

Finally, the total NoC power consumption (P_{NoC}) is represented by

$$P_{NoC} = P_R + P_L \qquad (2.10)$$

2.5.2 Area model

Before representing our area model, we should first emphasis that assuming a fixed basic infrastructure, as discussed at the beginning of Section 2.5, does not mean that the total utilized NoC area would be constant. Nevertheless, NoC utilized area would change depending on the number of links and the number of ports of the routers in the generated architecture. Accordingly, we represent our area model in this subsection for completeness. However, if the designer assumes a completely fixed infrastructure and wants to exclude this metric from the optimization process, this could be easily done using the weight factors represented in Subsection 2.5.5.

Similar to the NoC power, the NoC area consists of the area of its routers and links. As explained in Subsection 2.5.1, we are using a library of pre-designed routers with different number of ports. Accordingly, the total *routers area*, in μm^2, is expressed as

$$A_R = \sum_{i=1}^{N} A_{Ri} \qquad (2.11)$$

where N is the number of routers in the network, and A_{Ri} is the area of router i.

The link area depends on the number of wires per each link (i.e., the channel width), the wire length and width, and the spacing between wires. In this chapter, we assume a fixed channel width. Moreover, the wire width and spacing are technology-dependent and could be obtained from the corresponding technology library [250]. As a result, the *links area*, in μm^2, is expressed as in [146]

$$A_L = N_L \cdot (N_C \cdot (w_w + s_w) + s_w) \cdot l_l \tag{2.12}$$

where A_L is the links area, N_L is the number of links within the network, and N_C is the channel width. w_w, s_w, and l_l are the wire width, the inter-wires spacing, and the wire length for global interconnects, respectively.

Finally, the total NoC area (A_{NoC}), in μm^2, is expressed as

$$A_{NoC} = A_R + A_L \tag{2.13}$$

2.5.3 Delay model

The traffic from any source core experiences three types of delays in its way to the destination core. These are the arbitration and propagation delays through routers, the propagation delay through links, and the serialization and de-serialization delays through Network Interfaces (NIs). In this chapter, we use the average delay model represented in [239]. This model represents the delay in seconds. However, other models that give the delay as the number of clock cycles, like those in [89, 334], could be also used with our methodology. Given our assumptions, the overall NoC average delay (D_{NoC}), in *seconds*, from [239] is approximated as

$$D_{NoC} = \mu \cdot (t_a + t_r) + (\mu + 1) \cdot t_l + \frac{N_P}{N_C} \cdot t_l \tag{2.14}$$

where

$$t_a = (\frac{21}{4} \cdot log_2 p + \frac{14}{12} + 9) \cdot \tau \tag{2.15}$$

$$t_{l(r)} = 0.377 \cdot r_{l(r)} \cdot c_{l(r)} \cdot l_{l(r)}^2 + 0.693 \cdot (R_{d0} \cdot C_0 \atop + R_{d0} \cdot c_{l(r)} \cdot l_{l(r)} + r_{l(r)} \cdot l_{l(r)} \cdot C_{g0}) \tag{2.16}$$

$$l_r = 2 \cdot (w_r + s_r) \cdot N_C \cdot p \tag{2.17}$$

$$R_{d0} = 0.98 \cdot \frac{V_{dd}}{I_{dn0}} \tag{2.18}$$

where μ is the *average internode distance* [234] (i.e., average number of routers between a source node and a destination node). t_a and t_r are the router arbitration and propagation delays, respectively. Similarly, t_l is the link propagation delay. N_P and N_C are the number of bits per packet and the channel width, respectively. p is the average number of ports per router, and τ is the delay of a minimum sized inverter of the target technology. $r_{l(r)}$ and $c_{l(r)}$ are the per unit length resistance and capacitance of the link (router) wires, respectively. $l_{l(r)}$ is the wire length for the link (router cross-bar). R_{d0} and C_{g0} are the equivalent output resistance and the gate capacitance of a minimum sized inverter of the target technology, respectively. C_0 is the total input capacitance of a minimum sized inverter of the target technology, which is the summation of the gate and the drain capacitances. w_r and s_r are the wire width and the inter-wires spacing for router internal interconnects, respectively. Finally, V_{dd} is the supply voltage and I_{d0} is the drain current when both the drain and the gate voltages are equal to the supply voltage. The values of all the technology-dependent parameters could be obtained from the corresponding technology library [250]. Moreover, the average internode distance (μ), in *hops*, could be calculated using the connectivity matrix (C) of the architecture and the traffic distribution matrix (Λ) of the application as

$$\mu = \frac{\sum_{i=1}^{N} \sum_{j=1}^{N} \lambda_{ij} \cdot C_{ij}}{\sum_{i=1}^{N} \sum_{j=1}^{N} \lambda_{ij}} \tag{2.19}$$

2.5.4 Reliability model

NoC reliability calculation becomes currently dominated by the many noise sources existing in modern deep sub-micron technology [30]. Accordingly, in this chapter, we define the reliability of an NoC link as the probability of transmitting a packet successfully over that link in the presence of noise. Moreover, Gaussian model is used to represent the on-chip noise sources [87, 90]. As a result, the probability ($\mathcal{P}_{ij}$) of sending λ_{ij} packets over a single link (l_{ij}) is represented as

$$\mathcal{P}_{ij} = \left(1 - \frac{1}{\sqrt{2\pi}} \int_{\frac{V_{sw}}{2\sigma}}^{\infty} e^{-\frac{\mu^2}{2}} d\mu \right)^{\lambda_{ij}} \tag{2.20}$$

where V_{sw} is the voltage swing, and σ is the noise standard deviation. The overall NoC reliability is the probability of transferring all the packets from all the sources to all the destination in the presence of noise. Consequently, this overall NoC reliability (R_{NoC}) is represented as [90]

$$R_{NoC} = \prod_{i=1}^{N} \prod_{j=1}^{N} \mathcal{P}_{ij}^{c_{ij}} \tag{2.21}$$

where N is the number of cores within the application. $\mathcal{P}_{ij}$ is the probability of transmitting λ_{ij} packets successfully over link l_{ij} in the presence of noise and c_{ij} the number of links a packet goes through during its transition from the source core c_i to the destination core c_j, i.e., the number of hops.

2.5.5 Optimization objective function formulation

In this subsection, we formulate the optimization *objective function* by combining the cost and performance models represented in the previous subsections. For a given application, this function is used by our GA-based technique to generate the best NoC architecture. In general, it is required by this formulation to maximize the reliability while minimizing all other metrics. Therefore, our multi-objective function is formulated as the product of the power, the area, and the average delay over the network reliability. Using the product for the formulation ensures fairness between different metrics. This could be checked by differentiating the objective function to calculate the relative change in its value. This relative change is proportional to the summation of relative changes of all the metrics. Accordingly, our product-based formulation guarantees that non of the metrics will dominate the optimization process. Therefore, our objective function is expressed as

$$ f = \frac{(P_{NoC})^{\alpha_P} \cdot (A_{NoC})^{\alpha_A} \cdot (D_{NoC})^{\alpha_D}}{(R_{NoC})^{\alpha_R}} \qquad (2.22) $$

where α_P, α_A, α_d, and α_R are power, area, delay, and reliability weight factors, respectively, which could be used by the designer to control the optimization process. A higher value of any of these factors over the others gives high importance during the optimization to minimize the metric it represents compared to other metrics. Moreover, setting any of these factors to zero cancels any effect of the corresponding metric on the optimization process. Finally, P_{NoC}, A_{NoC}, D_{NoC}, and R_{NoC} are the NoC power, area, average delay, and reliability as expressed by (2.10), (2.13), (2.14), and (2.21), respectively.

2.6 Custom architecture generation using GA

This section presents our GA-based technique that generates the best custom NoC *architecture* with respect to the four metrics explained in Section 2.5. The inputs to our technique are the application core graph, the technology-dependent parameters, the optimization weight factors (α_P, α_A, α_d, and α_R), the frequency of operation (f), and the data resulted from analyzing the employed router library. Before running the GA, every router in the library is analyzed to find its area (A_R), its power constants (C_{DM}, C_D, C_{LM}, and C_L), and its maximum allowable Flit Arrival Rate (FAR). FAR is used to avoid any flit loss and to ensure that the generated architecture does not violate the maximum allowable bandwidth for routers. Finally, all routers in the library are checked to constitute the maximum allowable number of Ports Per Router (PPR). PPR is used to avoid generating an architecture with routers that are not available in the designer library.

As explained in Section 2.3, applying GA requires representing different NoC architectures in the form of chromosomes. Accordingly, our GA-based technique represents these architectures as binary vectors. As shown in Figure 2.3, a vector is a direct representation of the adjacency matrix (A) of that architecture. As we assume bidirectional links (i.e., $A_{ij} = A_{ji}$), it is actually sufficient to use either the upper or the lower triangle of the matrix. Therefore, the size of each vector is $N \cdot (N-1)/2$, where N is the number of cores in the application.

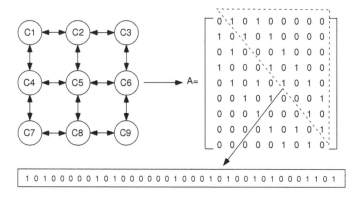

FIGURE 2.3
Example of a chromosome representation for 3×3 mesh architecture.

The MATLAB® GA toolbox [197] is used in solving the optimization problem in this chapter. Different selection, crossover, and mutation functions could be used with our methodology. However, for the results presented in this chapter, the toolbox is set to the following configurations:

- Creation function (Uniform): The initial population is created randomly with a uniform distribution.

- Selection function (Stochastic uniform): Every individual has the same chance of being selected. Although this selection function ignores the individual fitness, it has been chosen because of its suitability for small population size [187].

- Crossover function (Scattered): Offspring genes are chosen randomly from the two parents. Scattered crossover mixes the parents genes completely and allows for evolving good individuals [52].

- Mutation function (Uniform): This is the only built-in function for binary chromosomes in MATLAB®. Parents from the current generation are selected randomly and some of their genes are flipped. The number of genes to be flipped and their locations are chosen randomly.

2.6.1 Legality criteria for generated architectures

Architectures are generated randomly by the GA engine. Therefore, in this subsection, we present the *legality criteria* for any generated architecture. Any architecture that fails to meet certain legality constraints is considered invalid. In this chapter, a penalization technique[2] is employed to prevent the survival of invalid architectures

---

[2]Trying a restoration technique instead of the penalization one is left as a future work.

from one generation to the next. Accordingly, the optimization objective function (f) is multiplied by an architecture legality factor (β_l). This factor is 1 for any valid architecture, whereas a high legality factor is used for invalid architectures to increase the overall values of their fitness functions. The exact value of this legality factor should not be a problem as long as it is high enough to ensure that the value of the objective function of an invalid architecture is greater than that of any valid one. This requires the designer to estimate the worst case objective function value of valid architectures. If this value could not be estimated, a legality factor of more than 100 is proved, by experimentation, to be sufficient enough. Finally, the fitness function (F) of our GA-based technique is expressed as

$$F = f \cdot \beta_l \tag{2.23}$$

The fitness function (F) is used to evaluate different architectures. Any architecture is considered valid ($\beta_l = 1$) only if it meets the following constraints:

1. **Number of ports per router constraint**: This ensures that the generated architecture does not include a router with a number of ports greater than the PPR resulted from the router library analysis. Every row i in the adjacency matrix (A) expresses the connection of the corresponding router i to other routers in the architecture. Therefore, this constraint is formulated mathematically as

$$(\sum_{j=1}^{N} A_{ij}) + 1 \leq PPR, \quad \forall (i = 1 \; to \; N) \tag{2.24}$$

2. **Traffic continuity constraint**: This ensures that there is a path in the generated architecture between any two cores communicating with each other in the core graph. Accordingly, for any traffic (λ_{ij}), in the traffic distribution matrix, that is greater than zero, there should be a corresponding finite value (C_{ij}) in the connectivity matrix, i.e.,

$$\forall (\lambda_{ij} > 0) \implies C_{ij} = finite \tag{2.25}$$

3. **Flit arrival rate constraint**: This ensures that the flit arrival rate for any router in the generated architecture is less than its corresponding FAR, which is obtained from the router library analysis. Mathematically, this constraint is formulated as

$$\alpha_{f_{imax}} \leq FAR_i, \quad \forall (i = 1 \; to \; N) \tag{2.26}$$

where $\alpha_{f_{imax}}$ is the maximum flit arrival rate over all the ports of router i. Different flit arrival rates are calculated in our GA-based technique by routing all the traffic through the generated architecture according to the employed routing protocol. Therefore, flit arrival rate represents the amount of traffic per port.

2.6.2   Methodology for custom architecture generation

Figure 2.4 shows our methodology for custom architecture generation using GA. Before running the GA engine, area and power analysis is carried out on the employed router library to generate the required data for the GA optimization engine. This analysis results in different routers area, power constants, and maximum possible flit arrival rates. Moreover, it also indicates the maximum usable number of ports per router. The GA optimization starts, in the first step, by reading the inputs required by the engine. Thereafter, the GA engine runs iteratively to find the best architecture with respect to power, area, delay, and reliability. Each generation consists of 10 different chromosomes representing 10 different architectures. Every chromosome in any generation is evaluated by steps 2–10 in the flowchart. In the second step, the corresponding adjacency matrix of the chromosome is generated. The PPR constraint is then checked in the third step according to (2.24). The connectivity matrix for any architecture that passes the PPR constraint is generated in the fourth step by Dijkstra's algorithm. In the fifth step, the traffic continuity constraint is checked according to (2.25). Consequently, packets are routed from all the sources to all the destinations in the sixth step. Moreover, all input and output ports traffic is calculated for every router in the architectures. Based on this traffic, the flit arrival rate for all the ports is calculated in the seventh step. Consequently, the flit arrival constraint is checked in the eighth step according to (2.26). In the ninth step, the power, area, delay, and reliability of the architecture are calculated according to (2.10), (2.13), (2.14), and (2.21), respectively. These values are then used in the tenth step to calculate the GA fitness function according to (2.23). Any architecture that fails any of the constraints is marked invalid to prevent it from surviving to the next generation. In the eleventh step, once the exit criterion is satisfied, the GA engine generate the best architecture. As this architecture is not guaranteed to be free of loops so far,[3] the routing tables for all the routers in the architecture are generated to avoid any deadlock. Finally, if the exist criterion is not satisfied and a new generation is to be produced, the two architectures with the best fitness function are allowed to survive. Moreover, four new architectures are generated by crossover. Parents are selected randomly from the current generation and crossovered according to the employed MATLAB® configurations, as mentioned in Section 2.6. Finally, the four remaining architectures in each generation are produced by random mutation using the mutation function mentioned in Section 2.6.

Different values for population, elitism, crossover, and mutation could be used with our methodology. However, it is not recommended to use a high population size. The convergence time is a key parameter for ANoCs. High population sizes were found to increase the required running time of the optimizer significantly. For example, we tried using a population size of 100 chromosomes, which, on average, results in tripling the running time. This increase in the running time is caused by two things. First, the running time for the GA depends on the population size and the calculations required for every architecture within a generation. From one hand, using more architectures in each generation allows for reaching the optimum quickly.

---

[3]Proposing a routing algorithm and adding deadlock-free constraints are left as future work.

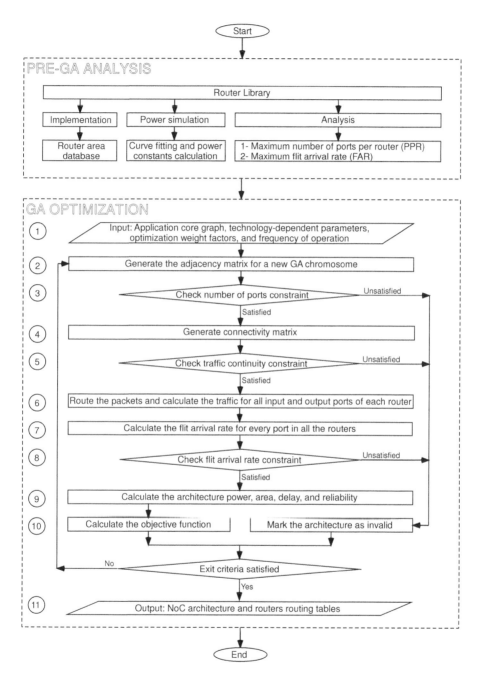

FIGURE 2.4
Methodology for custom architecture generation using GA.

On the other hand, every architecture requires time-consuming calculations of its power, area, delay, reliability, and overall objective function. The GA running time is found to be more proportional to these calculations than to the number of architectures within a generation. In other words, the reduction in the running time by using more architectures within a population is shadowed by the calculation overhead for these architectures. Second, an N-core application can have up to $N!$ different architectures. Unfortunately, most of these architectures are disconnected and hence are invalid. Consequently, increasing the population size causes more invalid architectures to be introduced every generation. This, in turn, causes the GA to lose a significant amount of time doing calculations for invalid architectures rather than converging to the optimum one.

2.7   Experimental results

We apply our GA-based technique on four real applications with different number of cores: Audio Video (AV) benchmark with 18 cores [85], Video Object Plane Decoder (VOPD) benchmark with 16 cores [274], MPEG-4 decoder with 12 cores [221], and a Multi Window Display (MWD) benchmark with 9 cores [85]. For every benchmark, we use our GA-based technique to generate five different architectures. These are the P architecture that is optimized solely for power ($\alpha_P = 1, \alpha_A = \alpha_D = \alpha_R = 0$), the A architecture that is optimized solely for area ($\alpha_A = 1, \alpha_P = \alpha_D = \alpha_R = 0$), the D architecture that is optimized solely for delay ($\alpha_D = 1, \alpha_A = \alpha_P = \alpha_R = 0$), the R architecture that is optimized solely for reliability ($\alpha_R = 1, \alpha_A = \alpha_P = \alpha_D = 0$), and the F architecture that is optimized for the overall objective function ($\alpha_P = \alpha_A = \alpha_D = \alpha_R = 1$). As an example, Figure 2.5 shows the five generated architectures for the AV benchmark. The core graph of the benchmark is shown in Figure 2.1. From these experiments, we notice:

1. For the P architecture, router power is proportional to the number of ports and the traffic per each port. Accordingly, the GA needs to keep the number of ports for each router as minimum as possible. Moreover, routing the traffic traces through the generated architecture should minimize the traffic per port. Finally, traffic paths should be kept as short as possible to minimize the link power. These three goals are conflicting. For example, minimizing the number of ports results in more traffic per port and longer traffic paths. However, our GA technique achieves the best tradeoff between these goals and generates the architecture with the minimum possible power. This is realized by optimally minimizing the number of ports and routing low traffic traces over longer paths of routers with low traffic.

2. For the A architecture, we experimented with both 90nm and 180nm technologies. For both of them, routers area dominates links area. Moreover, router area is directly proportional to the number of ports. Accordingly, the GA tries

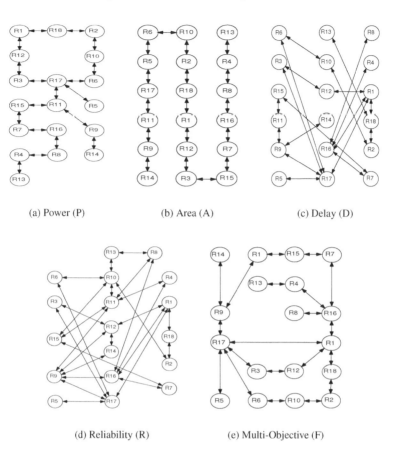

(a) Power (P)   (b) Area (A)   (c) Delay (D)

(d) Reliability (R)   (e) Multi-Objective (F)

FIGURE 2.5
Optimized architectures for (a) power, (b) area, (c) delay, (d) reliability, and (e) overall objective function for the AV benchmark.

to minimize the number of ports per each router. Many architectures result in the same minimum area. The one in Figure 2.5(b) represents the minimum-delay mapping of the cores onto what we might call a "line" architecture.

3. For the D architecture, there is a similar compromise to that for the P architecture. On one hand, it is required to minimize the number of ports to minimize the arbitration delay. On the other hand, traffic traces need more ports to go through the shortest paths to minimize the links and routers propagation delay. GA again achieves the best compromise between these conflicting goals. Interestingly, traces with low traffic are optimally routed over longer paths and through low traffic routers to minimize the overall NoC delay.

4. For the R architecture, maximizing the reliability requires that all traffic traces

are routed through the shortest paths to minimize the possibility of being affected by the on-chip noise. Consequently, the generated architecture is just a replica of the application core graph.

5. For the F architecture, our GA technique trades off all the previous four metrics and generates a multi-objective optimized architecture. As a result, the generated architecture achieves the best compromise among the independent design variables of these metrics. In a nutshell, these independent variables are the number of ports per router, traffic per ports, traffic routing, average number of hops, and the probability of the traffic being changed by noise.

To further explain the tradeoff between different metrics for different NoC applications, Figures 2.6 to 2.9 show the power, the area, the delay, the reliability, and the overall objective function for different architectures resulting from different generation techniques. P, A, D, R, and F, represent the architectures optimized for power, area, delay, reliability, and the four metrics together using our GA-based technique, respectively. Moreover, M represents a standard mesh architecture with application cores mapped by the NMAP algorithm [220]. Mesh is selected to represent standard architectures in the comparison as it is proved to provide the best tradeoff between different NoC metrics [237]. Similarly, NMAP is used for mapping because of its superior performance over other mapping techniques [220]. L and C are two architectures produced by semi-custom generation techniques. More precisely, L represents the architecture generated by the long-range link insertion methodology presented in [234]. C represents the architecture generated by the network clustering based technique presented in [211]. The former methodology customizes the architecture for the delay, whereas the latter aims at reducing the NoC area. We summarize our findings from these figures as follows:

1. All the architectures generated by our optimization-based technique, except for the A architecture, outperform standard and semi-custom architectures for all the applications.

2. Any single-metric optimized architecture outperforms other architectures for that metric. However, this single-objective optimization causes a clear deterioration in other metrics.

3. The A architecture specially gives bad results for any metric other than the area. Certainly, the use of the minimal possible resources results in a bad NoC performance.

4. The figures show a typical tradeoff between performance metrics (delay and reliability) and cost metrics (power and area). For example, the R architecture always results in high power and area. Similarly, the P and A architectures result in low reliability. As a matter of fact, increasing the reliability requires using more resources, which affects the power and the area badly.

5. From the multi-objective perspective, the F architecture results in the best tradeoff between all metrics. This is followed by the D and the P architectures.

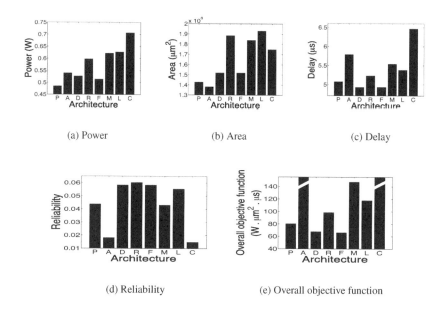

(a) Power   (b) Area   (c) Delay

(d) Reliability   (e) Overall objective function

FIGURE 2.6

Power, area, delay, reliability, and overall objective function comparisons of different architectures for the AV benchmark.

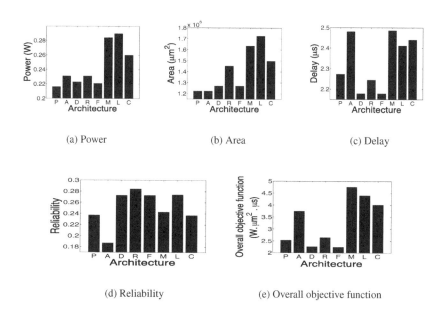

(a) Power   (b) Area   (c) Delay

(d) Reliability   (e) Overall objective function

FIGURE 2.7

Power, area, delay, reliability, and overall objective function comparisons of different architectures for the VOPD benchmark.

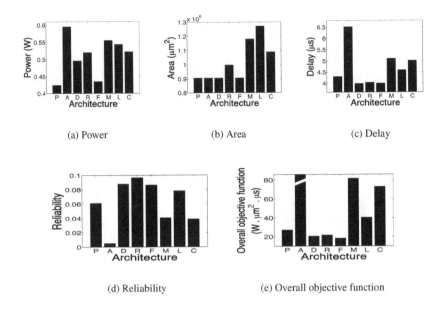

FIGURE 2.8

Power, area, delay, reliability, and overall objective function comparisons of different architectures for the MPEG-4 benchmark.

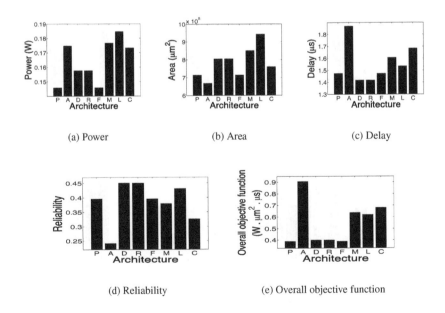

FIGURE 2.9

Power, area, delay, reliability, and overall objective function comparisons of different architectures for the MWD benchmark.

As our GA technique compromises between more design variables to generate these two architectures, they achieve better overall objective function than the R and the A architecture. The A architecture is affected by the great deterioration happened to its power, delay, and reliability. As a result, it sometimes scores even worse than the standard and the semi-custom architectures.

To further quantify our results, Table 2.2 shows the percentage enhancement of the F architecture over all other architectures for different metrics. The values in the table are the average over the four benchmarks. Negative values indicate that the corresponding architecture is better than the F architecture for the selected metric. This is usually corresponding to the architecture that is solely optimized for this metric. For example, power-wise, the F architecture is worse than the P architecture by 2.5%, whereas it outperforms the A, D, R, M, L, and C architectures by 16.5%, 6.4%, 12.2%, 24.6%, 26.2%, and 23.5%, respectively. Furthermore, the last column in the table represents the mean of these average values. For any metric, this represents the percentage enhancement of the F architecture over all other architectures over the four benchmarks. Accordingly, in average, the F architecture is better than the other architectures by 13.4%, 10.6%, 10.1%, 16%, and 209.1% for the power, the area, the delay, the reliability, and the overall objective function, respectively.

TABLE 2.2
Average percentage enhancement of the F architecture over other architectures for the four benchmarks

Metric	Average enhancement (%)							Mean
	Architecture							
	P	A	D	R	M	L	C	
Power	−2.5	16.5	6.4	12.2	24.6	26.2	23.5	13.4
Area	−2.4	−4.8	3.2	15.3	24.8	33.9	14.9	10.6
Delay	3.7	30.2	−1.0	1.6	15.8	9.7	20.7	10.1
Reliability	16.8	58.5	−3.9	−8.3	23.5	1.3	40.1	16.0
Overall objective function	20.9	1080.3	4.6	22.4	162.0	89.3	293.5	209.1

Table 2.2 also shows a strong correlation between the two performance metrics: delay and reliability. Architectures that are customized to lower the delay result in high reliability, and vice versa. Furthermore, the D architecture outperforms the F architecture for the reliability metric by 3.9%. Although the R architecture could not outperform the F architecture for the delay metric, it is only 1.6% worse than it. This could be explained by looking into the independent design variables that are included in both delay and reliability optimizations. From one hand, all design variables, i.e., number of hops and traffic routing, which dominate the reliability optimization are included in the delay optimization. Moreover, the F architecture is generated by trading off more design variables than the D architecture. These are the variables corresponding to the power and the area metrics. Accordingly, this enables the D architecture to outperform the F architecture with respect to the reliability. On

the other hand, some design variables that dominate the delay optimization, like the number of ports per router, are not included in the reliability optimization. Therefore, the R architecture could not outperform the F architecture with respect to the delay. Finally, a similar correlation could by noticed from the table between the cost metrics: power and area. Moreover, the power optimization includes all the design variables required for the area optimization. Consequently, the P architecture is more area-efficient than the F architecture.

Table 2.3 shows the average number of generations of our GA-based technique. The number of generations of any GA-based technique might differ from one trial to another based on the random generation of the chromosomes. Therefore, for every architecture, we ran the GA seven times. We further excluded the two trials with the maximum and the minimum values. We then averaged the remaining five trials. The table shows that the number of generations required to produce any of the five architectures is proportional to the number of cores within the application. Furthermore, for any benchmark application, the number of generations depends on the number of design variables to compromise in between during the architecture generation. Therefore, for a single-objective optimization, the generation of the P and the D architectures takes more longer than the A and the R architectures because the former trades off more design variables than the latter.

TABLE 2.3
Average number of generations for our GA-based technique

Application	Average number of generations				
	P	A	D	R	F
AV (18 cores)	125	58	77	40	396
VOPD (16 cores)	108	46	58	11	300
MPEG-4 (12 cores)	49	21	25	12	201
MWD (9 cores)	47	7	16	8	114

Finally, Table 2.4 shows the average running time of our GA-based technique, which corresponds to the average number of generations in Table 2.3. For our experiments, we used a desktop computer with 3.6 GHz Intel Xeon CPU and 2.75 GB RAM. Furthermore, the running times of two previous architecture optimization techniques for applications with the same number of cores are also included in the table. MILP is the mixed integer linear programming technique presented in [284], whereas ISIS is the GA-based technique presented in [282]. The maximum running time for our technique, for the F architecture of the AV benchmark, is 72.35 seconds. Accordingly, our technique outperforms previous optimization techniques with respect to the running time complexity. For example, the MILP technique failed to converge to a solution within a time out period of 8 hours for three of the benchmarks. Moreover, our methodology is at least 14 times faster than the ISIS technique. This low running time complexity of our technique makes it more attractive to ANoCs. Finally, a large amount of the running time of our technique is spent to generate the routing tables. As a result, the timing complexity could be further enhanced by using more elaborate adaptive routing strategies.

TABLE 2.4
Average running time for different architecture optimization techniques

Application	Average running time (seconds)						
	P	A	D	R	F	MILP	ISIS
AV (18 cores)	24.42	11.45	15.28	7.99	72.35	t.o.	≻1567
VOPD (16 cores)	20.03	8.63	10.79	2.43	58.23	t.o.	1564
MPEG-4 (12 cores)	9.23	4.09	4.77	2.63	30.27	t.o.	901
MWD (9 cores)	8.10	1.56	3.09	1.81	21.92	118	324

2.8 Conclusions

In this chapter, we presented a GA-based multi-objective technique for autonomous on-chip network architecture optimization. Our technique considered four NoC metrics: power, area, delay, and reliability. The models of the four metrics were discussed. The formulation of our fitness function was then presented. Finally, the GA-representation of the problem was explained. The optimization could be carried out for power, area, delay, reliability, or the four of them according to weight factors supplied by the designer. Results showed that the proposed technique is an efficient way to compromise between different NoC metrics. Moreover, the running time of our technique makes it more suitable for ANoC than previous architecture optimization techniques.

We plan to extend this work on three directions. First, we will try to enhance our reliability model to better represent other degradation factors, like link and router failures. Second, to enhance the running time of our technique, we will propose a routing algorithm to be used with it instead of the time consuming process of routing table generation. Third, we will integrate our technique with other ANoC monitoring and reconfiguration modules to implement a complete autonomous on-chip network.

3

An Autonomic NoC Architecture Using Heuristic Technique for Virtual-Channel Sharing

K. Latif

University of Turku, Finland
Turku Centre for Computer Science (TUCS), Finland

A. M. Rahmani

University of Turku, Finland
Turku Centre for Computer Science (TUCS), Finland

T. Seceleanu

ABB Corporate Research, Västerås, Sweden
Mälardalen University, Västerås, Sweden

H. Tenhunen

University of Turku, Finland
Turku Centre for Computer Science (TUCS), Finland

CONTENTS

Buffers are greatly influential for the overall network on chip based systems operation. The existence of buffers contributes to enhance the throughput performance but a big fraction of power in interconnection platforms is consumed by buffers. Therefore an intelligent, self-protected and fault tolerant architecture for buffer management in Network-on-Chip (NoC) is required to meet the modern system requirements. Virtual channel (VC) architecture improves the network performance by increasing the physical channel utilization and avoiding deadlocks. The buffer utilization in typical VC architecture still needs to be addressed. On other hand, fully shared buffer architectures can deliver a high throughput with higher degree of buffer utilization at the expense of power consumption and silicon area.

This chapter introduces a novel architecture for an ideal tradeoff between throughput, area and power. The *Partial Virtual Channel Sharing NoC* (PVS-NoC) architecture has been presented to overcome the issue of resource utilization and power consumption. Autonomic power saving, communication resource retrieval during faulty conditions, and reduction in crossbar size are the major contributions.

3.1   Introduction

Ever-increasing requirements on electronic systems are one of the key factors for evolution of the integrated circuit technology. Multiprocessing is the solution to meet the requirements of upcoming applications. Multiprocessing over heterogeneous functional units require efficient on-chip communication [30]. Network-on-Chip is a general purpose on-chip communication concept that offers high throughput, which is the basic support in dealing with the complexity of modern systems. All links in NoC can be simultaneously used for data transmission, which provides a high level of parallelism and makes it attractive to replace the typical communication architectures like shared buses or point-to-point dedicated wires. Apart from the high throughput, a NoC platform is scalable and has the potential to to keep up with the pace of technology advances [142]. However, all these features come at the expense of area and power. In the RAW multiprocessor system with 16 tiles of processing elements, interconnection network consumes 36% of the total chip power while each router dissipates 40% of individual tile power [317].

A typical NoC system consists of processing elements (PEs), network interfaces (NIs), routers and channels. The router further contains switch and buffers. Buffers consume the largest fraction of dynamic and leakage power of the NoC node (router + link) [20, 59]. Storing a packet in buffer consumes far more power as compared with its transmission [328]. Thus, increasing the utilization of buffers and reduction in number and size of buffers with efficient autonomic control enhances the system performance and reduces the area and power consumption.

Wormhole flow control has been proposed to reduce the buffer requirements and enhance the system throughput. But one packet may occupy several intermediate switches at the same time. In typical NoC architectures, when a packet occupies a

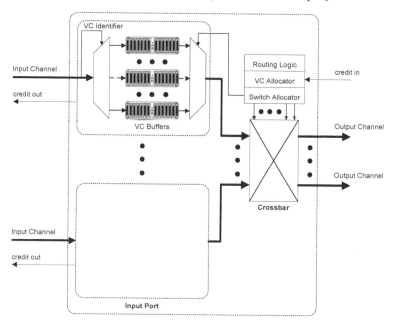

FIGURE 3.1
Typical virtual channel router architecture.

buffer for a channel, the physical channel cannot be used by other channels, even when the original message is blocked [30]. This introduces the problem of deadlock and livelock in wormhole scheme.

VCs are used to avoid deadlock and livelock. Figure3.1 shows a typical virtual channel router architecture [217]. Virtual channel flow control exploits an array of buffers at each input port. By allocating different packets to each of these buffers, flits from multiple packets may be sent in an interleaved manner over a single physical channel. This improves the throughput and reduce the average packet latency by allowing blocked packets to be bypassed. By inserting the VC buffers, we increase the physical channel utilization but the utilization of inserted VC buffers is not considered.

Now, the issue that needs attention is the utilization of both, the physical as well as the virtual channel utilization. A well designed network exploits available resources to improve performance [19]. ViChaR architecture [229] points out the VC buffer utilization, discussed later in Section 3.2. Even for ViChaR architecture, it can be observed that if there is no communication on some physical channel at some time instant, and the neighboring channel is overloaded, free buffers of one physical channel cannot contribute for congestion control by sharing the load of neighboring channel. Adaptive routing technique provides a solution to these issues but introduces some other problems like packet reordering.

Buffer utilization can be enhanced by sharing with the other ports autonomically, while not used by the current port. Fully shared buffer architectures can deliver a high throughput at the expense of area and power consumption with maximum degree of buffer utilization. Alternatively, typical NoC architectures – without any sharing of buffers – require less power to operate but throughput is also affected negatively. The buffer utilization is not good as well. If a buffer is available to receive the data but there is no data on input port, the buffer cannot be used by neighboring overloaded ports. Thus, each architecture has its own advantages and disadvantages, and an ideal tradeoff among performance, power, and area is necessary in order to obtain generally optimal operating systems.

Autonomic system-on-chip (ASoC) aims to solve the complexity and reliability issues of modern on-chip designs [332]. Autonomous architecture can optimize the NoC performance and reduce the power consumption to overcome the issues discussed above. Autonomic power gating can help to significantly reduce the static power consumption. In addition, autonomous channel allocation and routing can optimize the system performance. While autonomous routing may introduce some other issues, like packet reordering and deadlocks, autonomous buffer allocation can enhance the system performance, according to the specific system status.

Forecast-based dynamic virtual channel management has been proposed by [256], whit the purpose to reduce the power consumption. The approach uses the network traffic as well as past link and total virtual channel utilizations and dynamically forecasts the number of virtual channels that should be activated. It is based on an exponential smoothing forecasting method that filters out short-term traffic fluctuations. In this technique, for low (high) traffic loads, a small (large) number of VCs are allocated to the corresponding input channel. An autonomic power gating technique on the basis of forecasting approach can show a significant reduction in power consumption for shared virtual channel buffers between multiple ports because there are more options to turn off the virtual channel buffers due to enhanced buffer utilization

Other NoC architectural aspects relate to fault tolerance topics. Certain autonomic NoC platforms are fault tolerant and can deliver the desired performance without a large impact in case of certain manufacturing faults, by using the intelligent routing algorithms like [84, 98, 188, 333]. The key problem with these fault tolerant techniques and routing algorithms is that many resources become useless for the system when there is a fault on one resource. Due to the link failure in typical VC based NoC architectures, the VC buffers connected to it become useless for the system. To reduce the effect of fault on the system performance, such unused resources should be utilized by the system.

A well designed network exploits available resources to improve performance [19]. In this direction, we propose here a router architecture with enhanced utilization of VC buffers aimed at reducing the system latency, its power consumption, and silicon area. The chapter provides a novel architecture with autonomic sharing of buffers between two input ports with enhanced utilization of resources and reduced crossbar size and power consumption.

The remainder of this chapter is organized as follows. Section 3.2 presents an overview of the state of the art in NoCs especially using the VCs. Section 3.3 presents the link load analysis for NoC based system. Section 3.4 explains the proposed architecture, its architectural comparison with modern NoC architectures and the VC utilization in presence of faulty links. Finally, experimental results are presented, and we draw the conclusions.

3.2   Background

Several approaches have been proposed to enhance the buffer utilization in NoC based system. This comes as a performance improvement to utilize the unused buffers. This section reviews several relevant studies on this topic.

Lan et al. [167] addresses the buffer utilization by making the channels bidirectional and shows significant improvement in system performance. But in this case, each channel controller has two additional tasks: dynamically configuring the channel direction and allocating the channel to one of the routers, sharing the channel. Also, there is a 40% area overhead over the typical NoC router architecture due to double crossbar design and control logic. The extra tasks for channel controller and increased crossbar size contribute to the power consumption as well.

Nicopoulos et al. [229] presents a Dynamic *Virtual Channel Regulator* (ViChaR) for NoC routers. The authors address the buffer utilization by using the unified buffered structure (UBS) instead of individual and statically partitioned FIFO buffers. It provides each individual router port with a variable number of VCs according to the traffic load. The architecture provides around 25% improvement in system performance over an equal size generic router buffer at a small cost of power consumption. The architecture enhances the buffer utilization under heavy traffic load at the port. If there is no load at the port, the buffer resources cannot be utilized by neighboring overloaded ports.

Soteriou et al. [281] introduces a distributed shared buffer (DSB) NoC router architecture. The proposed architecture shows a significant improvement in throughput at the expense of area and power consumption due to extra crossbar and complex arbitration scheme.

Coenen et al. [68] presents an algorithm to optimize size of decoupling buffers in network interfaces. The buffer size is proportional to the maximum difference between the number of words produced and the number of words consumed at any point in time. This approach shows significant improvement in power dissipation and silicon area. The buffer utilization in idle time with optimal buffer size can contribute to reduce the overall system power consumption without affecting the system performance. If some buffer is idle at some time instant, it can share the load of neighboring input channels and thus increase the utilization of existing resources with a small control logic.

Neishabouri et al. propose the *Reliability Aware Virtual Channel* (RAVC) router architecture [225] which allocates more memory to the busy channels and less to the idle channels. This dynamic allocation of storage shows 7.1% and 3.1% latency decrease under uniform and transpose traffic patterns respectively at the expense of complex memory control logic. Though this solution is latency efficient, it is not area and power efficient.

The main motivation of this work is to propose a NoC architecture with considerations of all the issues discussed above. A NoC architecture with an optimal tradeoff among the throughput performance, power consumption and silicon area guides to propose the partial sharing of VC buffers.

3.3   Resource utilization analysis

To enhance the utilization of interconnection resources, it is necessary to make the analysis of each network resource individually. The traffic pattern for general purpose MPSoC architectures is usually unpredictable, when multiple applications run simultaneously. Routing algorithm controls the utilization of communication channels. The system utilizes the router resources according to the load on incoming channels. Thus, the link load analysis can provide the base to formulate the solution for utilization of communication resources. The bandwidth requirements for communication and task scheduling are known for NoCs, when designed for specific set of applications like NoCs for multimedia applications. This helps to optimize the resource utilization.

3.3.1   Link load analysis

In synthetic traffic analysis, the average load for each link is calculated with a variety of routing algorithms. As a test case, uniform, transpose, bit complement and NED traffic patterns were analyzed with XY routing. In the uniform traffic pattern, a node sends a packet to any other node with an equal probability while in transpose traffic pattern, each node (i,j) communicates only with node (j,i). For bit complement traffic load, each node (i,j) communicates only with node $(M-i,N-j)$, if mesh size is MxN. The NED is a synthetic traffic model based on Negative Exponential Distribution where the likelihood that a node sends a packet to another node exponentially decreases with the hop distance between the two cores. This synthetic traffic profile accurately captures key statistical behavior of realistic traces of communication among the nodes [254]. Figure 3.2 shows the percentage load for each link on the network for different traffic patterns measured by Eq. 3.1.

$$L_{(i,j)\to(k,l)} = \frac{\text{TLL: } (i,j) \to (k,l)}{\text{TNL}} \quad\quad (3.1)$$

where

$$\text{TLL: } (i,j) \to (k,l) = \sum_{x=0}^{x=M}\sum_{y=0}^{y=N} (S(x,y),D(k,l)|via(i,j))$$

and

$$\text{TNL} = \sum_{\substack{0<m<(M-1)\ 0<n<(N-1)}} \sum_{\substack{0<o<(M-1)\ 0<p<(N-1)}} L_{(m,n)\to(o,p)}$$

To measure the total link load (TLL) on a specific link directed from node (i, j) towards (k, l), the traffic load from the sources represented by $S(x, y)$ routed via node (i, j) towards destination node $D(k, l)$ is considered. For total network link load (TNL), the link load from all the interconnection links is summed up. The expression is topology independent and can be extended to any number of dimensions.

Now consider the node "12" in Figure3.2(b). The input ports to receive data from nodes "11" and "13" are not used at all during the whole simulation independent of the total simulation time. But the input ports from left and right receive the traffic load. The traffic load from node "22" is double than the load from node "02". In case of typical NoC architectures, the link between nodes "12" and "22" is overloaded all the time but cannot utilize the available resources of other ports. Similar behavior can be observed for odd-even routing.

For any router, two ports are overloaded compared with other two for all the traffic patterns. In most cases, the identical traffic load value is for another input port as well. Thus, there are only two load values, and each value is for two ports of the same router. Another interesting observation is that the traffic load values are different for opposite directions. For load balancing and resource utilization, the resources can be shared among a pair of input ports. The pair of ports is selected in such a way that one port has higher load value and other one has a lower load value. The resources can be shared among all the input ports, but this increases the input cross bar size, which, in turn, increases the power consumption, area and average packet latency due to the more complex necessary controller. Another important and interesting issue is that sharing the resources among two ports with balanced loads can deliver the degree of resource utilization and the throughput closer to the *Fully Virtual-Channel Shared NoC* (FVS-NoC) architecture with significant reduction in power consumption.

3.3.2   Real application-based analysis

The MPEG4 application presented by [154] was selected for resource utilization analysis. The application with bandwidth requirements is shown in Figure3.3. Now consider the link loads for DR-SDRAM node. The right and down side ports share the router resources, while the left and up side ports also share the resources. The heavy load value 942 MB/sec from right port can utilize the resources from down port, which contains a smaller load value of 60.5 MB/sec. Thus, sharing communication resources among two ports can balance the input load on all ports while significantly increasing the crossbar size. By using this approach, the average load on input ports will be similar to the full sharing of resources.

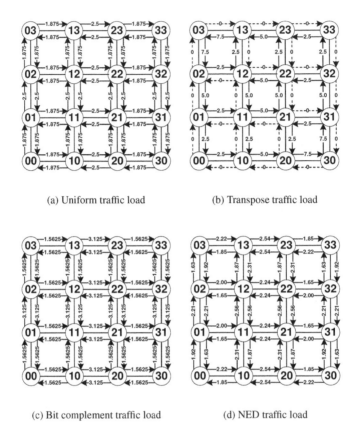

(a) Uniform traffic load   (b) Transpose traffic load

(c) Bit complement traffic load   (d) NED traffic load

FIGURE 3.2
Traffic load analysis for XY-routing.

3.4   The proposed router architecture: PVS-NoC

To enhance the performance of typical VC architectures, new VC buffers should be inserted because a congested port cannot utilize the resources of neighboring free port. The network can be fully utilized by sharing the resources among all the input ports. This increases the control logic complexity, crossbar size and power consumption, hence a tradeoff between resource utilization and power consumption is needed. Instead of sharing the VC buffers among all the input ports, if the buffers are shared autonomously among two ports, the buffer utilization will be increased and will bring the utilization closer to the fully shared architecture, as discussed in Section 3.3. This, without a big power consumption and a large area overhead. The number of buffers

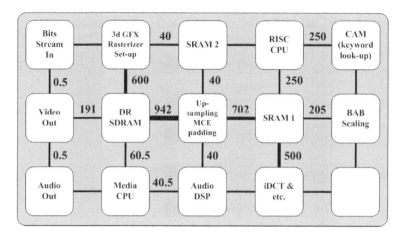

FIGURE 3.3
MPEG4 application.

can be reduced because of enhance utilization. This reduces the power consumption significantly because the router buffers consume the major fraction of power and area for NoC interconnection as discussed in Section 3.1. Thus, a partially shared virtual-channel NoC (PSV-NoC) has been proposed as tradeoff between resource utilization and power consumption.

The PSV-NoC architecture uses a two-stage control logic. At the input, it uses a hierarchy of arbiters, while the central control logic is used for output crossbar. The system level PVS-NoC architecture is shown in Figure3.4. The input crossbars and control logic are responsible for autonomic buffer allocation and receiving the data packets as explained in Section 3.4.2. The output crossbar and control logic use the typical NoC out architecture and its tasks are route computation and packet transmission as discussed in Section 3.4.3. Both control logics operate according to the VC buffer status and do not communicate with each other. The detailed PSV-NoC architecture is shown in Figure 3.5. The signaling details for each component are explained later in this section.

The need for effective plug-and-play design styles pushed the development of standard interface sockets, allowing to decouple the development of computational units from that of communication architectures [153]. The same needs are applicable to our developments, too, and we solve the issues with the help of wrappers. A wrapper, thus, provides the abstraction between PE and the interconnection platform. The wrapper architecture can be divided into two parts: the Network Interface (NI) and the PE Interface (PI) as shown in Figure 3.6. NI is a generic interface, which needs to be standardized. Main tasks of the NI are packetization and de-packetization of data. Different services can be introduced in NI architecture, such as multicasting and error monitoring.

The queuing buffer for the PE can be considered as the part of NI on input port of

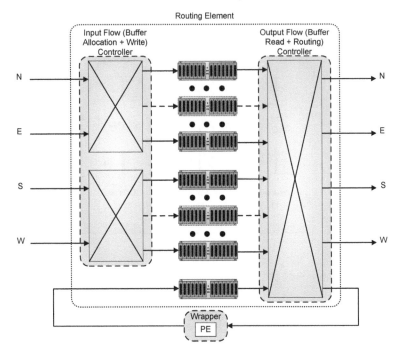

FIGURE 3.4
System level specification of router architecture.

the switch. Thus each PE has its own dedicated buffer, which simplifies the control logic and enhance the throughput without any area overhead. PI is the core specific interface, which acts as a clock synchronizer. Different services can be introduced in the PI architecture, such as thermal monitoring and power safe mode, while the core is not in use. Due to this abstraction and synchronization, the NoC platform can be seen as a plug and play platform. This helps to reduce the design time, and heterogenous PEs can be integrated to the system.

3.4.1 Packet format

Data are packetized by the NI and it is injected into the network according to the format shown in Figure 3.7. The header flit contains the source address (SA), the operational code (OP), the priority level (PR) and the destination address (DA):

SA. This field contains the address of the packet source (or request initiator for control signal communication). Devices on the platform are identified by unique numbers, which are used for addressing as well.

OP. The *OP* signal is used to identify the *Operation* to be done on the packet, if destination node PE offers multiple operations.

PR. The *PR* field is used for buffer allocation, when only one buffer is available and both neighboring routers are requesting the buffer allocation or in similar situation

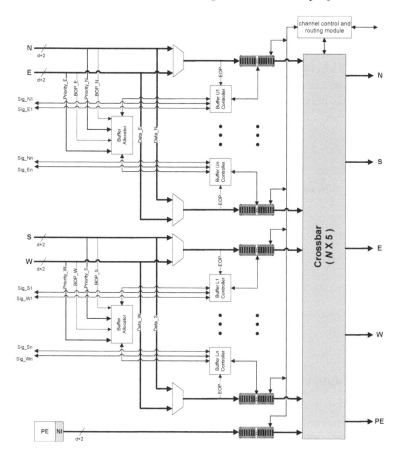

FIGURE 3.5
System level specification of router architecture.

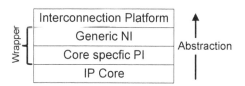

FIGURE 3.6
NoC wrapper structure.

for output port allocation, the router with packet of higher *Priority* level wins the allocation.

DA. This field contains the address of the targeted device.

BOP. Beginning of Packet (BOP) identifies the header flit. Buffer allocation unit works on this signal and allocates the appropriate buffer. Output port allocator also works on BOP signal and computes the path according to routing policy to allocate the output port.

EOP. End of Packet (EOP) signal is used to mark the allocated buffer as free and available for next allocation. Similarly, Output port controller knows that packet transmission has been completed. EOP is also a mark for tail flit.

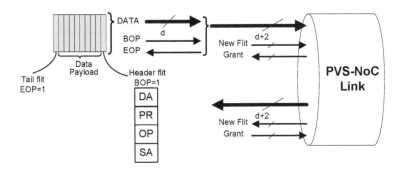

FIGURE 3.7
Packet format.

3.4.2   The input controller and buffer allocation

Most of the PVS-NoC contribution stands on the input side because the issue is to enhance the buffer utilization by allocating the free buffers to overloaded ports, which is the task of the input control logic. The VC buffers are shared between neighboring input ports as shown in Figure3.5. The input architecture of two ports is replicated for other pairs of input ports and can be extended to any number of input ports according to the topology requirements. The PE uses the dedicated buffer for packet injection into the network. The detailed architecture of the input side for a pair of neighboring ports is shown in Figure3.8.

The buffer dedicated to the local PE is not shared with any other port. In this section, two input ports are used to explain the complete operation. The control of two ports which share the VC buffers is completely independent from the other input control logic as shown in Figure3.5. The task of the *Buffer allocator* is to keep track of free buffers and allocate them to the incoming traffic. Independent control logic belongs to each *Buffer allocator*. After allocation, the *Buffer write* control directly communicates with the neighboring nodes. This distributed control reduces the communication overhead and power consumption. The input control system operates in following phases:

Buffer Allocation: The *Buffer allocator* receives the requests from neighboring routers for new buffer allocation in the form of *BOP* signal. With the status of all the buffers known, the allocator allocates the free buffers to the requesting router. If only one buffer is available and both routers are requesting for the buffer, the router

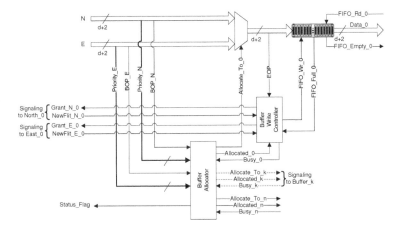

FIGURE 3.8
Architecture of input part of router.

with higher priority value (*PR*) in header flit is allowed to use the buffer for packet transmission.

Local Signaling: After the allocation, the corresponding *Buffer write* controller is informed by raising the *Allocated* signal from *Buffer allocator* to the *Buffer write* controller. In response, the *Buffer write* controller raises the *Busy* signal. At the same time, the *Buffer allocator* signals the corresponding multiplexer by *Allocated_To* signal, to which input, the corresponding buffer has been allotted.

Packet Receive: After local control signaling, the *Buffer write* controller signals the *Grant* to the corresponding neighbor router. The latter uses the *Grant* signal as VC identifier for requested packet and always raises the equivalent *NewFlit* signal, while transmitting flits for that packet.

Buffer De-Allocation: The *Busy* signal from the *Buffer write* controller goes down upon arrival of tail flit marked - *EOP* signal. After receiving *Not Busy* signal from the *Buffer write* controller, the *Buffer allocator* marks the corresponding buffer as free; the *Status_Flag* is raised.

Status Flag: The *Status_Flag* signal is the logical *AND* operation of all the *Busy* signals from *Buffer write* controllers. When all buffers have been allocated, the *Status_Flag* is raised to inform the neighboring routers that no buffer is available for transmission of new packets. If at least one buffer is available for allocation, the *Status_Flag* is down.

The *buffer allocation* unit works only when the *BOP* signal is received. Once the buffer has been tied to the requesting router, the *Buffer allocator* goes into the sleep mode to save power. It does not consume power until the next BOP signal is received. Similarly, the *Buffer write* controller works only after receiving the *Allocated* signal from *Buffer allocator* and goes in sleep mode at the *EOP* signal.

3.4.3   The output controller and routing algorithm

The *Output* part is modeled by a typical N×5 crossbar, where N is the total number of buffers in the router including the buffer dedicated to local PE. The internal architecture of the output crossbar is shown in Figure 3.9. The crossbar size can be customized according to the topology requirements. The mesh topology was considered to decide the number of I/O ports of the router. Distributed control logic was used here as well. There is one *central controller*, which is the key part of the router. The central output controller decides the routing policy and computes the route for the packet. The port controller controls the packet transmission and the communicates with the destination router, without involving the *central controller*. It also decides the flow control as well. A worm-hole flow control was used, which makes efficient use of buffer space as small number of flit buffers per VC are required [73]. The Output controller operation can be divided into several steps, as follows.

Route Computation: The *central controller* senses the buffers and, on *BOP* signal, computes the route for the packet by using *DA* field as explained in Section 3.4.1.

Local Signaling: After computing the route, the corresponding output port is assigned to the packet. Then, the *central controller* informs the corresponding port controller by the signal *New_Buffer_Allocation* which means that the new packet has been allocated on this port for transmission. In parallel, the *central controller* sends the buffer ID containing the packet to the port controller and FIFO read logic block. After that, the *central controller* marks that buffer ID as *allocated* internally. It does not read the *BOP* signal for a buffer which is marked *allocated*.

Data Read and Transmission on Output Ports: The *Port controller* generates the read signal for that buffer. If the FIFO empty signal is high, the *port controller* does not generate the read signal for that buffer. The FIFO read logic is required to avoid multiple drivers problem for buffer reading. The FIFO read controller keeps track of allocation and allows only the appropriate port to read the buffer.

De-Allocation: When the *EOP* bit is high, the port controller marks that buffer as free and sends the corresponding buffer ID back to the *central controller* to mark it as *unallocated*. After that, the *central controller* starts checking the *BOP* bit for route computation and portal allocation.

Communication Between Port Controller and Central Controller: Communication signaling between the *central controller* (routing element) and the port controller is shown in Figure3.10. The FIFO read logic is a combinational logic module intended to avoid multiple driver error. The *central controller* computes the route for the incoming header flit and sends the corresponding output port ID to the port controller by the signal *Buffer_Allocation_ID_N*. In parallel, the *central controller* raises the signal *New_Buffer_Allocation_N* to inform the port controller that a new buffer has been allocated to it for transmission. After transmitting the tail flit, the *port controller* sends back the *Buffer ID* for deallocation by the signal *Buffer_Deallocation_ID_N*. After that, *central controller* is waiting for next BOP signal from that buffer for next output port assignment to incoming packets. *Port_Status_N* is raised, when corresponding port cannot make more assignments.

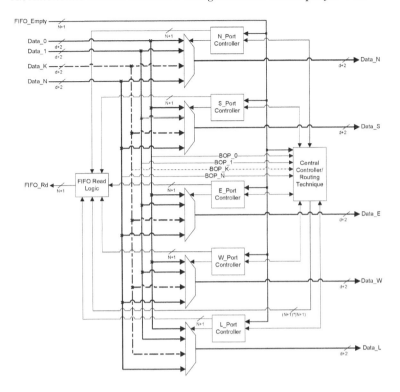

FIGURE 3.9
Architecture of output part of router.

3.4.4 Comparison with existing architectures

To make the comparison with existing architectures, 10-port router with 2 ports per direction was selected with 10 internal buffers. Table. 3.1 shows a comparison of PVS-NoC architecture with other NoC architectures.

The typical VC NoC represents a conventional virtual channel NoC architecture with 2 VCs per port and use unidirectional channels to communicate with neighboring routers. Thus two channels are required between two neighboring routers for two way communication. In case of heavy traffic load on a certain port, typical VC architecture can provide only 2 VCs to receive the packets on that port. The PVS-NoC can provide 4 VCs to the same port under heavy traffic load on same port. Thus, the VC utilization has been doubled, at the price of a slight overhead in crossbar size.

BiNoC has two bidirectional channels, which can be used according to the requirements to communicate in any direction [167]. Compared with the BiNoC architecture with 10 in-out channels, the PVS-NoC approach provides 5 input and 5 output channels. PVS-NoC can provide 4 input and 4 output VCs for bidirectional communication between two nodes. However, BiNoC provides only two physical channels for packet transfer between two nodes, in any direction. The option of di-

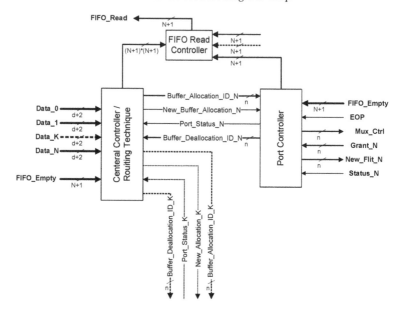

FIGURE 3.10
Output control architecture.

rection selection is provided at the cost of a large crossbar size. Another issue to be addressed here is scalability. To insert a new buffer in BiNoC architecture, a separate *Buffer allocator* is required and cross bar size will increase exponentially with increase in number of I/Os. For PVS architecture, the number of VC buffers can be selected according to the application and topology requirements for the proposed architecture. To insert a new VC in PVS design, the buffer and a controller are needed without any modification in existing logic but slight increase in crossbar resources.

The DSB-175 and DSB-300 router architectures have been detailed in [281]. To make an exact comparison, we define a DSB-160 architecture following the descriptions in [281]. The DSB-160 is the router with 160 flits of aggregate buffering. The buffers are divided between 5 middle memory banks with 16-flit buffers per bank and aggregate 80-flit input buffers comprising two 8-flits buffers (VCs) at each input port. Five memory banks are not considered in comparison table 3.1 because only one flit can be written into and read from a middle memory in DSB architecture, which reduce the utilization of memory banks. Thus the static power consumption without increasing the system performance is the major overhead of DSB architecture as compared with the PVS-NoC.

In FVS-NoC architecture, any of 10 VC buffers can be allocated to any input port to receive the traffic. All the channels are unidirectional. The architecture provides the maximum utilization of VC buffers at the cost of extra big crossbar, which makes the solution area and power expensive as compared with our PVS-NoC.

TABLE 3.1

Comparison with existing NoC router architectures

Architecture⇒ Resource⇓	Typical VC NoC	BiNoC	Distribute Shared Buffer NoC (DSB-160)	FVS-NoC	PVS-NoC
Number of Buffers	10	10	10	10	10
Channels/Direction	1-in 1-out	2-inout	1-in 1-out*	1-in 1-out	1-in 1-out
Maximum VCₛ/ Physical Channel	2	1	2	10	4
Buffer Size	16 flits	16 flits	8 flits	16 flits	16 flits
Total Buffer Size	160 flits	160 flits	160 flits	160 flits	160 flits
Crossbar	10X5	10X10	2(5X5)	5X10 and 10X5	2(2X4) and 10X5

* 5 memory banks are not considered as only one flit can be written into and read from a middle memory.

3.4.5 Virtual-channel sharing under faults

The additional feature of PVS-NoC is to retain the system performance till certain level after the occurrence of fault. In NoC based interconnection platform, the fault can occur in three types of components: physical link, buffer and the controller.

Consider here that a fault has occurred on a network inter-router-link. In typical architectures, the input/output buffers connected to the link cannot be utilized in future. Suppose further that there are "X" faulty links in NoC based system and each input port contains "V" virtual channels with buffer depth 'd' and each flits size or buffer width is "f" bits. If there is a small VC controller for each virtual channel, the resources useless after the fault on physical channel can be estimated by Eq. 3.2.

$$Resource\ Overhead = X \cdot V \cdot d \cdot f_{(Memory\ Cells)} + X_{(VC\ Control\ Logic)} \\ + X \cdot V \cdot f_{(wires\ in\ Output\ Crossbar)} \quad (3.2)$$

Power gating techniques can be applied to save the static power consumption for resources mentioned in Eq.3.2. This means, however, that the "gated-out" resources are no longer available for the network. If these buffers could still be used by the NoC system, the fault will not impact the system performance significantly and routing intelligence can help to retain the performance.

Consider the NoC platform shown in Figure3.11. For node "11" as source and node "13" as destination, the route of packet will be just to go up vertically in normal situation. If the communication link between nodes "11" and "12" is broken, the packet will be rerouted as shown. Similarly, when node "10" is the source and node "30" is the destination. Now, the resources on the new routes and especially the router at node "21" will be overloaded. The input ports from left and down side will be overloaded due to the faulty links, while other inputs will operate with normal loads in typical architectures. To balance the load on input buffers and provide a relief to the loaded ports, if the right and down input ports share the VC buffers and left and upper input ports as well, the extra load due to faulty links will be shared by all the ports. PVS-NoC architecture uses this sharing approach to retain the performance in case of fault on any communication resource.

To retain the performance after the fault occurrence, the problem needs to be

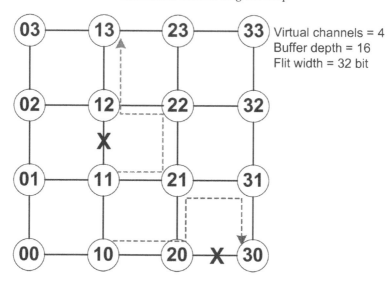

FIGURE 3.11
Different routes between two source-destination pairs in presence of faulty links.

approached from a different way. Fault on physical link also makes the buffer and controller useless for the system. In case of PVS-NoC architecture, all the VC buffers and controllers can be used by the neighboring port. This reduces the impact of fault on system performance. In case of fault on buffer, the load on corresponding port can be shared by neighboring port and impact of faulty buffer on port load becomes exactly half for the PVS-NoC architecture and congestion can be avoided. In case of fault in control logic, none of available architectures can utilize the rest of resources related to the fault link.

3.5   Experimental results

To demonstrate the performance characteristic of the PVS-NoC architecture, a cycle-accurate NoC simulation environment was implemented in VHDL. The packets had a fix length of seven flits, the buffer size was eight flits and the data width was set to 32 bits. The 5 × 5 2D mesh topology was used for interconnection. With same parameters, typical virtual channel and FVS-NoC architectures were analyzed. Static XY routing was used. The proposed architecture was analyzed for synthetic and real application traffic patterns.

3.5.1   Synthetic traffic

In synthetic traffic analysis, the performance of the network was evaluated using latency curves as a function of the packet injection rate. The packet latency was defined as the time duration from when the first flit is created at the source node to when the last flit is delivered to the destination node. For each simulation, the packet latencies were averaged over 50,000 packets. Latencies were not collected for the first 5,000 cycles to allow the network to stabilize. To perform the simulations, XY wormhole routing algorithm [HU03a] under uniform, transpose and Negative Exponential Distribution (NED) [10] traffic patterns was used.

The throughput curves for uniform, transpose and NED traffic patterns are shown in Figure3.12. It can be observed for all the traffic patterns, the PVS-NoC architecture saturates at higher injection rates as compared with the typical VC architecture but slightly less than FVS-NoC architecture. The reason is that partial VC sharing makes the load balanced. In the case of PVS-NoC, bandwidth limitations are managed by proper resource utilization without increasing the communication resources. The saturation point of the PVS-NoC architecture is just before FVS-NoC because FVS-NoC provides more buffer utilization by sharing the VC buffer among all input ports. However, FVS-NoC is not a power efficient solution as verified for realistic application traffic pattern.

3.5.2   Real application traffic

For realistic traffic analysis, the encoding part of video conference application with sub-applications of H.264 encoder, MP3 encoder and OFDM transmitter was used. The application graph with 25 nodes is shown in Figure3.13. The application graph consists of processes and data flows; data is, however, organized in packets. Processes transform input data packets into output ones, whereas packet flows carry data from one process to another [169]. A transaction represents the sending of one data packet by one source process to another, target process, or towards the system output. A packet flow is a tuple of two values (P, T). The first value "P" represents the number of successive, same size transactions emitted by the same source, towards the same destination. The second value "T" is a relative ordering number among the (packet) flows in one given system. For simulation purposes, all possible software procedures are already mapped within the hardware devices. The application mapped to 5×5 2D-mesh NoC is shown in Figure3.14.

To estimate the power consumption, we adapted the high level NoC power simulator presented provided by [117]. The power consumption of the interconnection network (NoC switches, bus arbiters, intermediate buffers, and interconnects) is based on 35nm standard CMOS technology. The simulation results for average packet latency (APL) and power consumption for video conference encoding application are shown in Table.3.2. The PVS-NoC showed 18% reduction in power consumption but 6% more APL over the FVS-NoC architecture. On other hand, the PVS-NoC showed 22.32% reduction in APL value but 7.9% more power consumption over the typical VC architecture. Thus, the proposed architecture PVS-NoC provides an optimal tradeoff between APL and power consumption.

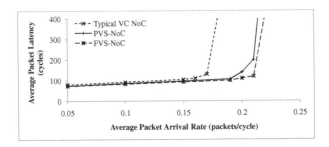

(a) Uniform traffic load.

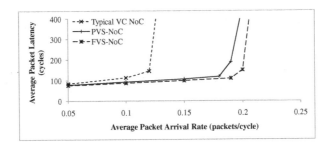

(b) Transpose traffic load.

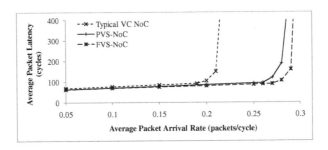

(c) NED traffic load.

FIGURE 3.12
Average packet latency (APL) vs. packet injection rate for 5×5 Mesh 2D NoC.

3.6 Conclusions

An autonomous PVS-NoC architecture where an ideal tradeoff between virtual channel utilization, system performance and power consumption was proposed. The vir-

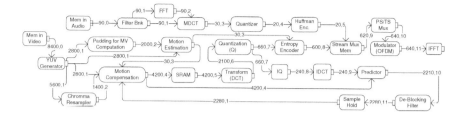

FIGURE 3.13
Video conference encoder application.

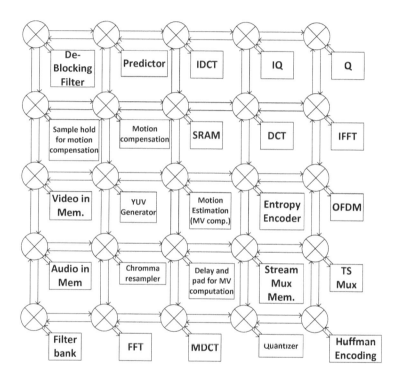

FIGURE 3.14
Video conference encoder application mapped to 5 × 5 Mesh NoC.

tual channel buffers are shared between two input ports in PVS-NoC. Apart from system performance, the architecture is also fault tolerant. In case of any link failure, the VC buffers are used by the other physical channel sharing the VC buffers and system performance is not affected severely.

The proposed architecture was simulated with synthetic and real application traffic patterns. The performance was compared with typical VC based NoC architecture

TABLE 3.2
Experimental result for power consumption and APL of video conference encoder
application running at 50 MHz

Architecture	Power Consumption (W)	APL (cycles)
Typical VC	66.4	112
PVS-NoC	72.1	87
FVS-NoC	87.9	82

and FVS-NoC. The PVS-NoC architecture showed significant improvement in system throughput without significant power consumption overhead.

Acknowledgments

The present work is supported by Nokia Foundation and the DOMES project funded by the Academy of Finland, project number 123518/2008.

3.7   Glossary

PVS-NoC: Partial Virtual-channel Shared NoC Architecture.

FVS-NoC: Fully Virtual-channel Shared NoC Architecture.

4

Evolutionary Design of Collective Communications on Wormhole NoCs

J. Jaros

Brno University of Technology, Czech Republic

V. Dvorak

Brno University of Technology, Czech Republic

CONTENTS

This chapter describes the technique of the evolutionary design aimed at scheduling of collective communications on autonomic networks-on-chip (ANoC). In order to avoid contention for links and associated delays, collective communications proceed in synchronized steps. A minimum number of steps is sought for the given network

69

topology and given sets of sender and receiver nodes. The proposed technique is not only able to re-invent optimum schedules for known symmetric topologies like hypercubes but also can find schedules even for any asymmetric, irregular, multistage and fat topologies in case of general many-to-many collective communications. In most cases, the number of steps reaches the theoretical lower bound for the given communication pattern; if it does not, non-minimum routing can provide further improvement. Optimal schedules may serve for writing high-performance communication routines for application-specific networks on chip or for the development of communication libraries in the case of general-purpose interconnection networks.

4.1 Introduction

A recent trend in high performance computing (HPC) has been towards the use of parallel processing to solve computationally-intensive problems. Nowadays, with the enormous transistor budgets of 45 nm and 32 nm technologies on a silicon die, it is feasible to place large CPU clusters on a single chip (System on Chip, SoC). The memory of many-core systems is physically distributed among computing nodes that communicate by sending data through a Network on Chip (NoC) [140]. With an increasing number of processor cores, memory modules and other hardware units in the latest chips, the importance of communication among them and of related interconnection networks is steadily growing.

Communication operations can be either point-to-point, with one source and one destination, or collective, with more than two participating processes. Some embedded parallel applications, like network or media processors, are characterized by independent data streams or by a small amount of inter-process communications [142]. However, many general-purpose parallel applications display a bulk synchronous behavior: the processing nodes access the network according to a global, structured communication pattern.

The performance of these collective communications (CC) has a dramatic impact on the overall efficiency of parallel processing. The most efficient way to switch messages through the network connecting multiple processing elements makes use of pipelined wormhole (WH) switching [226]. Wormhole switching reduces the effect of path length on communication time, but if multiple messages exist in the network concurrently (as it happens in CCs), contention for communication links may be a source of congestion and waiting times. To avoid congestion delays, it is necessary to organize CC into separate steps in time and to put into each step only such pair-wise communications whose paths do not share any links. The contention-free scheduling of CCs is therefore important.

At present, many different universal topologies of interconnection networks are in use [45, 114, 263] and other application-specific ones can be created on demand [49, 149, 218]. While the lower bounds on time complexity of certain communication patterns can be mathematically estimated considering a particular network topology and a distribution of transmitters and receivers, finding a sequence of communication steps (a schedule) approaching this limit is an outstanding problem [57], and in some cases, such schedules are not known so far [126].

Naturally, many projects have addressed the design of fast collective communication algorithms for wormhole-switched systems in recent years. Since any data dropping is not acceptable in NoC, the deadlocks, livelocks and starvations, even links/nodes overloads, have to be prevented in such schedules. Hence, many approaches have analyzed the structure and properties of underlying NoC topology and communication pattern with the aim of designing contention-free communication schedules that attain the lower bound of time complexity of given CC patterns [23, 178, 266]. Unfortunately, these topology-specific schedules work for a few regular topologies like hypercube or square mesh and tours even then only in some possible instances. Another idea is to design some families of parametrized algorithms that can be tuned to perform well on different architectures under various system conditions [22]. Unfortunately, this kind of CC schedules is not optimal in most cases, and moreover they are restricted by other parameters of the NoC such as a port model, minimal routing strategy, symmetry of the network, etc.

With an increasing number of novel NoC topologies (e.g., Spidergon [149], Kautz [286], optimal diameter-degree topologies [218] and fat topologies) hunger for a general technique capable of producing optimal or near optimal schedules for an arbitrary network topology and the given CC pattern steadily grows. The designed schedules could serve for writing high-performance communication functions for a particular topology. Consequently, these functions could be included into, for example, the well-known OpenMPI library [302] to accelerate given CCs.

Due to the complexity of this task, evolutionary algorithms (EA) are employed in this work. Since EAs were introduced in 1960s [130], several researchers have demonstrated that EAs are effective and robust in handling a wide range of difficult real-world problems such as optimization, decomposition, scheduling and design [48, 60, 183].

In this chapter, we would like to experimentally confirm the following hypothesis [143]. *Evolutionary design is able to produce optimal or near-optimal communication schedules comparable or even better than those obtained by a conventional design for the networks of interest. Moreover, evolutionary design will reduce many drawbacks of present techniques and invent still unknown schedules for an arbitrary topology and scatter/broadcast communication patterns.*

4.2 Collective communications

A collective communication [163] is often the most efficient way to carry out a communication operation on a parallel computer. The reason is that if many nodes need to communicate with other nodes at the same point in the algorithm, a specialized communication operation in which the nodes cooperatively participate in the communication operation clearly has the opportunity for using the communication network more effectively, thus giving higher performance. Such specialized collective communication operations should thus be designed so that they cause as little overhead as possible.

4.2.1   A model of communication

Performance of CCs is closely related to the switching technique the network is operating on. Wormhole switching (WH) [226] is a special case of cut-through switching. Instead of storing a packet completely in a node and then transmitting it to the next node, wormhole switching operates by advancing the head of a packet directly from incoming to outgoing channels of the routing chip. A packet is divided into a number of flits (flow control digits) for transmission. The header flit (or flits) governs the route. As soon as a node examines the header flit(s) of a message, it selects the next channel on the route and begins forwarding flits down that channel. As the header advances along the specified route, the remaining flits follow in a pipeline fashion. Because most flits contain no routing information, the flits in a message must remain in contiguous channels of the network and cannot be interleaved with the flits of other messages. The time-space diagram of wormhole-switched message is shown in Figure 4.1.

Wormhole switching avoids memory bandwidth in the nodes through which messages are routed. Only a small FIFO (First-In-First-Out) flit buffer can be used. It also makes the network latency largely insensitive to path length. The blocking characteristics are very different from packet-switching. If the required output channel is busy, the message is blocked in place. The particular flits of a blocked message are stored along a part of a path from the source to the blocking node.

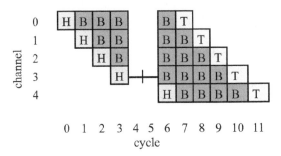

FIGURE 4.1

Time-space diagram of wormhole switching under contention.
(H - Head flit, B - body flit, T - tail flit).

The simplest time model of point-to-point communication in direct WH networks takes the communication time composed of a fixed start-up time t_s at the beginning (SW and HW overhead of a sender and a receiver), a serialization delay, i.e., transfer time of m message units (words or bytes), and of a component that is a function of distance h (the number of channels on the route or hops a message has to do):

$$t_{WH} = t_s + mt_1 + ht_r, \qquad (4.1)$$

where t_1 is per unit-message transfer time, and t_r includes a routing decision delay, switching and inter-router latency. A relatively small dependence on h may be taken into account by including $h_{max}t_r$ into t_s, so that only two components t_s and mt_1 are sufficient in the communication model.

In the rest of the chapter we assume that the CC in WH networks proceeds in synchronized steps. In one step of CC, a set of simultaneous packet transfers takes place along complete disjoint paths between source-destination node pairs. If the source and destination nodes are not adjacent, the messages go via some intermediate nodes, but processors in these nodes are not aware of it. The messages are routed automatically by the routers attached to processors.

Complexity of collective communication will be determined in terms of the number of communication steps or equivalently by the number of start-ups τ^{CC} (upper bound). Provided that term $h_{max}t_r$ is included in t_s and excluding contention for channels, CC times can be obtained approximately as the sum of start-up delays plus associated serialization delays $m_i t_1$ in individual communication steps

$$\tau_{CC} = \sum_{i=1}^{\tau^{CC}} (t_s + m_i t_i) = \tau^{CC}[t_s + mt_1]. \qquad (4.2)$$

The above expression assumes that the nodes can only retransmit or consume original messages, so that the length of messages $m_i = m$ remains constant in all communication steps. This is true in the so called non-combining model of communication; on the contrary, in the combining model the nodes can combine/extract partial messages with negligible overhead. The combining/non-combining model influences CC performance and either one can outperform the other in some cases. Further on we will consider only the non-combining model because it requires only a very simple switch architecture and small buffers.

Possible synchronization overhead involved in communication steps, be it hardware or software based, should be included in the start-up time t_s. Let us note that with uniform messages and a single clock signal domain, one barrier synchronization before CC might be sufficient to synchronize the whole CC. Communication steps would then follow in the lockstep. According to the frequency of CCs and an amount of interleaved computation (BSP model [35]) in a certain application, efficiency of parallel processing can be estimated.

The port model of the system defines the number k of CPU ports that can be engaged in communication simultaneously. This means that beside $2d$ network channels, there are $2k$ internal unidirectional (DMA) channels, k input and k output channels, connecting each local processor to its router that can transfer data simultaneously. Always $k \leq d$, where d is a node degree; a one-port model ($k = 1$) and an all-port router model ($k = d$) are most frequently used. Figure 4.2 shows a one-port and an all-port router for the network node with $d = 3$. In the one-port system, a node must transmit (and/or receive) messages sequentially. The messages may block on occupied injection channel, even when their required network channels are free. Architectures with multiple ports alleviate this bottleneck. There are as many local CPU channels as there are network channels. The all-port router architecture thus reduces the message blocking latency during collective communication operations and the port model determines the number of required start-ups and the CC performance.

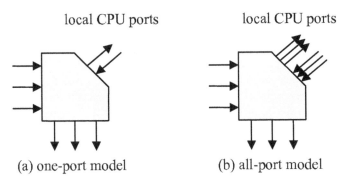

FIGURE 4.2
Port models for 3-regular networks.

4.2.2   Classification of collective communications

If V is the set of all processors, $|V| = P$, collective communications can involve communication among up to all P processors. Provided that there is 1:1 mapping between the set of processors (terminal nodes/processing elements) and processes, we can equivalently talk about communicating process groups. Generally we have two process groups: the subset of transmitters (senders) $T \subseteq V$ and the subset of receivers $R \subseteq V$. The subsets T and R can be overlapping and can be as large as V. We can distinguish three classes of CCs:

(1) $T \cap R = \emptyset$, non-overlapping sets of processes

  (a) One-to-All, $|T| = 1$, $|R| = P - 1$, e.g., One-to-All Broadcast (OAB) or One-to-All Scatter (OAS).

  (b) One-to-Many, $|T| = 1$, $|R| < P - 1$, e.g., Multicast (MC).

  (c) All-to-One, $|T| = P - 1$, $|R| = 1$, e.g., All-to-One Gather (AOG) or All-to-One Reduce (OAR).

  (d) Many-to-Many, $|T| = M$, $|R| = N$; $M, N < P$, e.g., non-overlapping sets of processes such as Many-to-Many Broadcast (MNB) or Many-to-Many Scatter (MNS).

(2) $0 < |T \cap R| < P$, Many-to-Many communication with overlapping sets of processes.

(3) $|T \cap R| = P$, All-to-All communications such as permutation, All-to-All Scatter (AAS), Broadcast (AAB), Reduce (AAR), etc.

In one-to-all CCs, one process is identified as a sender (called also the root or initiator), and all other processes in the group are receivers. Here we assume that the root is also a receiver, however, the transfer is not done by means of message sending, but only by local memory-to-memory copy. In all-to-all CCs, all processes in a process group perform their own one-to-all or all-to-one CCs. Thus, each process will receive $P - 1$ messages from $P - 1$ different senders in the process group. Finally, many-to-many collective CCs are the generalizations of all previously presented CCs. In these cases, only a subset of nodes can be involved by CCs.

Since complexities of some communications are similar (AOG $\sim$ OAS, AOR $\sim$ OAB, AAR $\sim$ AAB), we will focus only on scatter- and broadcast- based CCs. Scatter [21] is a CC pattern where a root distributes a unique message to each other process (nodes in the interconnection network). Broadcast [198] is a CC pattern where a root distributes the same message to all other processes. Scatter differs from broadcast in that the root node starts with $P - 1$ unique messages, one destined for each node. Unlike broadcast, scatter does not involve any duplication of data (see Figures 4.3 and 4.4).

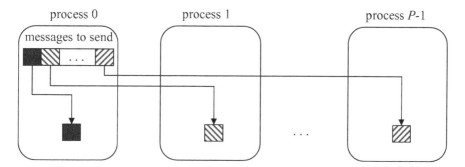

FIGURE 4.3
The basic idea of one-to-all scatter. Private messages are distributed.

4.2.3 The lower bounds on time complexity

One of the key design factors of an interconnection network is its topology. The lower bounds $\tau_{CC}(G)$ for the network graph G depend on node degree d, number of nodes P, and bisection width B_C, Table 4.1. Let us note that the lower bound cannot be exactly derived for irregular topologies because the node degree d may not be constant. The lower bound can be estimated only by taking into account the lowest and the highest node degree in such a topology [163, 300].

As far as the broadcast communication (OAB) is concerned, the lower bound on the number of steps $\tau_{OAB}(G) = s = \lceil \log_{k+1} P \rceil$ is given by the number of nodes

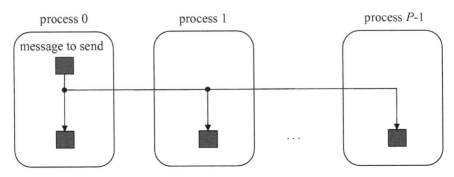

FIGURE 4.4
The basic idea of one-to-all broadcast. The same message is distributed.

TABLE 4.1
Lower complexity bounds of selected CCs (any regular topology)

CC	WH, k-port, FD links, non-combining model
OAB/AOR	$\lceil \log_{k+1} P \rceil = \lceil (\log P)/\log(k+1) \rceil$
AAB/AAR	$\max(\lceil \log_{k+1} P \rceil, \lceil (P-1)/k \rceil)$
OAS/AOG	$\lceil (P-1)/k \rceil$
AAS/AAG	$\max(\lceil (P^2/B_C) \rceil, \lceil \Sigma/(Pk) \rceil, \lceil (P-1)/k \rceil)$

informed in each step, that is initially $1, 1 + 1 \times k$ after the first step, $(k + 1) + (k + 1) \times k = (k+1)^2$ after the second step, ..., and $(k+1)^s \geq P$ nodes after step s. Since the broadcast message is the same for all the nodes, each node once informed can help with distributing of the message in following steps.

In case of AAB communication, since each node has to accept $P - 1$ distinct messages, the lower bound is $\lceil (P-1)/k \rceil$ steps. A similar bound applies to OAS communication, because each node cannot inject into the network more than k messages in one step.

The lower bound for AAS can be obtained considering that one half of messages from each processor cross the bisection, whereas the other half do not. There will be altogether $\lceil 2(P/2)(P/2)/B_C \rceil$ of such messages in both ways, where B_C is the channel bisection width [82]. Sometimes a stronger lower bound may be obtained considering the count of channels from all sources to all destinations (Σ) and the limited count Σ_1 of channels available for one step. In regular networks with constant node degree $\Sigma_1 = Pk$. As each node has to accept $P - 1$ distinct messages, $\lceil (P-1)/k \rceil$ bound has to be also obeyed. Which lower bound takes effect depends on a particular network topology and the port model.

4.3   State-of-the-art

This subsection summarizes known approaches and results in the area of scheduling of collective communications (designing the communication algorithms):

- The OAS problem can be solved very simply by implementing the Sequential Tree (ST) also called separate addressing [21].

- On the contrary, in case of AAS, the theoretical and empirical techniques fail. Of course, the Mixed Integer Linear Programming (MILP) [223] or graph coloring [103] can be used, but the optimal schedule does not have to be discovered under deterministic routing or time limitation.

- The problem with broadcasts is that each node already informed can become a distributor of the message. Optimal algorithms reaching the theoretical lower bound of steps are not known even for familiar hypercubes of higher dimensions ($d > 7$) [126]. But for more complex topologies than simple meshes/tori or low dimensional hypercubes, it is still an open problem.

- For AAB, the number of steps is limited by k messages that can be absorbed by any node in one step. Therefore, we can inform only adjacent nodes in one step and still develop optimum scheduling. The task is easier in symmetric networks: it is sufficient to find the Time-Arc Disjoint Tree (TADT) [300], which translated to all source nodes, creates no conflicts in any step of AAB communication. However, for asymmetric or irregular networks no similar systematic approaches exist.

- Many of the described methods are restricted by other limitations. We can mention, for example, port-model limitation, deterministic routing, one processor per node, and regularity of underlying network.

We have considered only one-to-all or all-to-all communications so far. What about many-to-many communications? This area is completely unexplored, perhaps except for one-to-many broadcast more often referred to as multicast. The importance of many-to-many communication expresses in huge parallel systems with thousands of processing nodes, where many tasks are executed simultaneously and each task utilizes only a subset of these nodes.

There are many opened problems related to many-to-many CC scheduling:

- Many-to-many communication patterns are insolvable by the most of proposed techniques, because they use a part of the network showing up irregular topology.

- Only nodes belonging to a given group can participate in a CC schedule and help, for example, with the distribution of messages; other nodes in the topology are not aware of it.

- Any fault in NoC makes the topology also irregular meaning that CC schedules are not solvable by known techniques.

- Connecting more processors (cores) to a single node router creates a fat node. Fat networks can be seen as a set of interconnected routers that forward the messages between attached processors and the interconnection network. Because routers cannot consume or produce messages, all the CCs on fat topologies have to be treated as many-to-many ones.

- A closely related problem is how to spatially distribute processes onto nodes to ensure the best performance.

Finally, as we can conclude that the systematic approach, applicable to an arbitrary topology and able to obtain an optimal or near optimal schedule for any scatter or broadcast CC pattern does not exist as yet. One of the reasons is that complexity of the scheduling problem proves to be NP-complete [57]. For this reason a novel technique has been developed and implemented to alleviate all presented restrictions and bottlenecks. Its description follows in the next section.

4.4   Evolutionary design of collective communications

Since the 1960s there has been an increasing interest in the metaphor of imitating the evolution of living beings to develop powerful algorithms for difficult optimization problems. The algorithms belong to the class of stochastic algorithms named EAs [130]. EAs are powerful, domain-independent search techniques inspired by Darwinian theory. Since EAs were introduced, several researchers have demonstrated that EAs are effective and robust in handling a wide range of difficult real-world problems such as optimization, decomposition, scheduling and design [48,60,183].

In general, EAs employ selection and recombination to generate new search points in a state space. A genetic algorithm [108], typical representative of EA, starts with a set of individuals that forms a population of the algorithm. Usually, the initial population is generated randomly using a uniform distribution. On every iteration of the algorithm, each individual is evaluated using the fitness function and the termination function is invoked to determine whether the termination criteria have been satisfied. The algorithm ends if acceptable solutions have been found or the computational resources have been spent. Otherwise, the individuals in the population are manipulated by applying different evolutionary operators such as mutation and crossover. Individuals from the previous population are called parents while those created by applying evolutionary operators to the parents are called offsprings. The consecutive process of replacement forms a new population for the next generation (see Figure 4.5).

A lot of special variants of EAs have been designed and used. A promising modifications are Estimation and Distribution Algorithms (EDA) [168]. In EDAs, there

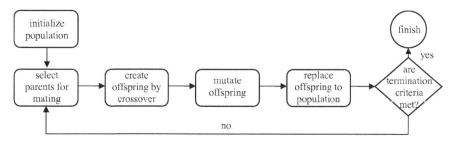

FIGURE 4.5
Flowchart of a standard genetic algorithm.

are neither crossover nor mutation operators. Instead, the new population of individuals is sampled from a probability distribution which is estimated from a database containing selected individuals from the previous generation. The advantages of EDA cover the ability of capturing linkage between genes and lower number of control parameters.

There are two important issues with respect to search strategies in EAs: exploiting the best solution and exploring the search space [42]. EAs provide a directed random search in complex landscapes. Recombination operators perform essentially a blind search; selection operators hopefully direct the search towards the desirable area of solution space. One general principle for developing an implementation of evolutionary algorithms for a particular real-world problem is to make a good balance between exploration and exploitation of the search space.

In general, a genetic algorithm has five basic components, as summarized by Michalewicz [199]:

(1) A genetic representation of solutions to the problem

(2) A way to create an initial population of solutions

(3) An evaluation function rating solutions in terms of their fitness

(4) Recombination operators that alter the genetic composition of children during reproduction

(5) Values for the parameters of genetic algorithms

In order to achieve a good balance between exploration and exploitation of the search space, all components of the evolutionary algorithms must be examined carefully. Additional heuristics should be incorporated in the algorithm to enhance the performance. The most important parts will be reviewed in the following subsections.

4.4.1 Input data structure and preprocessing

An input data structure maintains a description of a network topology, a definition of CC and sets of transmitters, receivers and intermediate routers. The topology description is saved in the form of nodeneighbor lists, where the nodes are considered to be neighbors only if they are connected by a simple direct link.

After the input file is loaded, the data have to be preprocessed. In the first phase, the preprocessor divides the set of all nodes V^* into a set of transmitters T and a set of receivers R. Then, a set of terminal nodes $V \subseteq V^*$ is determined as the union $T \cup R$. The terminal nodes can inject/consume messages to/from the network, while the non-terminal nodes (routers) can only retransmit the messages. Finally, all the sets are ordered based on the node index.

The preprocessor generates all the shortest paths (the set R_{xy}) between all transmitter–receiver pairs x-y and saves them into a specific data structure in the operating memory during the second phase. This task is performed by a modified well known Breadth-First Search (BFS) algorithm [106].

4.4.2 Scatter encoding

As broadcast and scatter are completely different communication services, candidate solutions are encoded in separate ways. First, the definition and features of scatter encoding will be introduced and examined.

Consider a scatter based CC communication between M transmitters from set T and N receivers from set R.

(1) The CC can be defined as a set COMM of pair-wise transfers src, dst originating in $src \in T$ and terminating in $dst \in R$, where $src \neq dst$.

$$COMM = \{comm_{src,dst} : src \in T, dst \in R, src \neq dst\} \qquad (4.3)$$

(2) A direct encoding can be designed for the scatter-based communication schedules (i.e., an exact description of the schedule is stored in a chromosome). A chromosome can be formalized as n-tuple of genes:

$$chromosome = \begin{pmatrix} gene_{0,0} & \cdots & gene_{0,N-1} \\ \vdots & \ddots & \vdots \\ gene_{M-1,0} & \cdots & gene_{M-1,N-1} \end{pmatrix} \qquad (4.4)$$

where M is the number of transmitters and N is the number of receivers while n is the total number of genes. Notice that

$$M, N \leq P, n = M \cdot N \qquad (4.5)$$

(3) A gene $gene_{i,j}$ represents a single message transfer from the transmitter (source) $x_i \in T$ to the receiver (destination) $x_j \in R$, where $x_i \neq x_j$. The source

and the destination are identified by the genes indexes i and j. A gene is the ordered couple:

$$gene_{i,j} = (l_{i,j}, s_{i,j}), 0 \leq i < M, 0 \leq j < N, i \neq j$$
$$l_{i,j} \in R_{i,j} \qquad (4.6)$$
$$0 \leq s_{i,j} < Steps$$

The first component $l_{i,j}$ represents a chosen path from the transmitter x_i to the receiver x_j stored in the set $R_{i,j}$. The second component $s_{i,j}$ determines a selected time slot (communication step) of the transfer between x_i and x_j. The total number of time slots (communication steps) is given by the predefined parameter *Steps*.

(4) The (shortest) path is defined as an ordered set of unidirectional channels:

$$l_{i,j} = \{c_1, c_2, c_3, \ldots, c_L\} \qquad (4.7)$$

where $src(c_1) = x_i$ and $dst(c_L) = x_j$.

(5) Next, consider a set *GENOME* containing all the genes included in a *chromosome*:

$$GENOME = \{gene_{i,j} : 0 \leq i < M, 0 \leq j < N, i \neq j\} \qquad (4.8)$$

(6) Finally, we can define bijective mapping f from set *GENOME* into set *COMM* meaning that each gene corresponds to a unique pair-wise transfer and also vice versa:

$$f : gene_{i,j} \in GENOME \mapsto comm_{src,dst} \in COMM \iff$$
$$x_i = src, x_j = dst \qquad (4.9)$$

4.4.3 Broadcast encoding

Consider a broadcast based CC communication between M transmitters from set T and N receivers from set R. Unfortunately, the set *COMM* cannot be constructed for broadcast based CCs in advance, since each node already informed could also become a distributor of the broadcast message.

(1) Therefore, the definition of CC is based on a set of messages *MSG* that have to be delivered during a given CC to each destination. Let

$$MSG = \{msg_{src,dst} : src \in T, dst \in R, src \neq dst\} \qquad (4.10)$$

be a set of messages originating in transmitters *src*, transported through the network via an intermediate distributor, and consumed by receivers *dst*. Notice that message distributors are not known in advance.

(2) A direct encoding has been designed to store broadcast-based communication schedules. A broadcast CC schedule is represented by the chromosome in the form of n-tuple of genes

$$chromosome = \begin{pmatrix} gene_{0,0} & \cdots & gene_{0,N-1} \\ \vdots & \ddots & \vdots \\ gene_{M-1,0} & \cdots & gene_{M-1,N-1} \end{pmatrix} \qquad (4.11)$$

where M is the number of transmitters, and N is the number of receivers, while n is the total number of genes. Notice that

$$M, N \leq P, n = M \cdot N \qquad (4.12)$$

(3) A gene $gene_{i,j}$ determines the way a receiver $x_j \in R$ obtains the broadcast message $msg_{i,j}$ from the transmitter $x_i \in T$ via a distributor $d_{i,j} \in V$. The producer and consumer of the message are identified by the genes indexes i and j. Individual genes are represented by ordered triplets:

$$\begin{aligned} &gene_{i,j} = (d_{i,j}, l_{i,j}, s_{i,j}), 0 \leq i < M, 0 \leq j < N, i \neq j \\ &d_{i,j} \in D_{i,s_{i,j}} \subset V \\ &l_{i,j} \in R_{d_{i,j},j} \\ &0 \leq s_{i,j} < Steps \end{aligned} \qquad (4.13)$$

The first component of the gene selects a distributor $d_{i,j}$ of the message $msg_{i,j}$. Besides the transmitter, the distributor can be any node from set $D_{i,s_{i,j}}$ that includes all nodes informed during all of $s_{i,j} - 1$ steps; see the extended domination set theory [297]. The second component $l_{i,j}$ represents a chosen path between the distributor $d_{i,j}$ and a receiver x_j from the set $R_{d_{i,j},j}$. Analogously, the last component $s_{i,j}$ determines a selected time slot of the transfer between $d_{i,j}$ and x_j. The total number of time slots (communication steps) is given by the predefined parameter $Steps$.

(4) The (shortest) path was defined in the same way as in Section 4.4.2 as an ordered set of unidirectional channels.

(5) Next, consider a set $GENOME$ containing all genes included in a *chromosome*:

$$GENOME = \{gene_{i,j} : 0 \leq i < M, 0 \leq j < N, i \neq j\} \qquad (4.14)$$

(6) Finally, we can define a bijective mapping f from set $GENOME$ into set MSG; thus each gene corresponds to a unique src-dst transfer of the message and also vice versa:

$$\begin{aligned} &f : gene_{i,j} \in GENOME \mapsto msg_{src,dst} \in MSG \iff \\ &x_i = src, x_j = dst \end{aligned} \qquad (4.15)$$

4.4.4   Fitness function definition

This section proposes a formal description of the fitness function. The definition is the same for scatter and broadcast CC.

(1) Let *SS* (Same Slot/Step) be a binary relation on the set *GENOME*. Let $a, b \in COMM$ (scatter) or $a, b \in MSG$ (broadcast) be message transfers represented by $gene_{i,j}$ and $gene_{k,l}$, then

$$gene_{i,j} SS gene_{k,l} \iff s_{i,j} = s_{k,l} \tag{4.16}$$

Thus, two transfers are in relation *SS* if and only if they are executed during the same time slot.

Now, we show that SS is an equivalence relation:

  (a) *SS* is reflexive, since no transfer can be performed in more than one time slot.

  (b) *SS* is clearly symmetric considering $s_{i,j} = s_{k,l}$, then $s_{k,l} = s_{i,j}$.

  (c) *SS* is transitive. Let a, b, c be elements of *GENOME*. Whenever $aSSb$ and $bSSc$, then also $aSSc$ (a is executed during the same slot as c whenever a is executed during the same slot as b and b is executed during the same slot as c).

Thus SS is an equivalence relation.

(2) The equivalence relation SS induces the partition on set *GENOME*. Each equivalence class $[g_s]$ includes all transfers performed in the same slot s.

(3) Let $E_{a,b}$ be a set of all channels shared in two transfers a, b represented by genes $gene_{i,j}, gene_{k,l} \in GENOME$ which utilize paths $l_{i,j}$ and $l_{k,l}$. Then

$$E_{a,b} = l_{i,j} \cap l_{k,l} \tag{4.17}$$

The number of conflicts between a and b can be obtained as the cardinality of the set $E_{a,b}$.

(4) Define a multiset E_s including channels shared by all transfers within a given time slot, then

$$E_s = \bigcup_{a,b \in [g_s], a \neq b} E_{a,b} \tag{4.18}$$

(5) The multiset E, covering all shared channels within the whole CC, can be obtained by a union over all equivalence classes. Thus

$$E_s = \bigcup_{[g_s] \in GENOME/SS} E_s \tag{4.19}$$

(6) The total number of conflicts can be obtained as the cardinality of multiset E. Thus

$$Fitness = |E| \qquad (4.20)$$

The valid communication schedule for a given number of communication steps must be conflict-free. Valid schedules are either optimal (the number of steps equals the lower bound) or suboptimal. Evolution of a valid schedule for the given number of steps is finished up as soon as fitness (number of conflicts) drops to zero. If it does not do so in a reasonable time, the prescribed number of steps must be increased.

4.4.5 Acceleration and restoration heuristics

New heuristics have been developed to improve OAS/AAS optimization speed taking into account a search space restriction due to a limited message injection capability of network nodes. Because no node can send more than k message in a communication step, an acceleration heuristic checks this condition in the whole chromosome and redesigns terminal node utilization in all communication steps before evaluating the fitness function. The implementation of the heuristic is based on a node utilization histogram.

The second OAS/AAS heuristic replaces the mutation operator in the employed EA. This heuristic swaps associated communication steps of two transfers originating in the same transmitter. Actually, this heuristic performs a local search on a candidate solution based on the time domain. The conflicting transfers are rescheduled in spite of their timing to reduce total congestion.

Since illegal solutions of the OAB/AAB problem can appear during the process of genetic manipulation with OAB/AAB chromosomes, (a gene violates the condition $d_{i,j} \in D_{i,s_{i,j}}$ in eq. 4.13), the restoration heuristic has to be proposed. This heuristic proceeds in subsequent communication steps and constructs a correct broadcast schedule. A check is made for every node whether the node receives the message really from the node already informed. If not so, the source node of this point-to-point communication is replaced by a node that has already received the message. The replacement is made taking into account the node utilization histogram. An exchange of the distributor node $d_{i,j}$ has naturally an impact on utilized links $l_{i,j}$. Hence the original path is replaced by the newly chosen one from a list $R_{d_{i,j},j}$ of exploitable paths between new input-output pair $d_{i,j}$ and x_j.

In order to accelerate the convergence of the EA, an OAB/AAB-specific heuristics have been developed. This heuristic injects good building blocks into the initial population. First, the communication step $s_{i,j}$ is initially set to the same value ($s_{i,j} = 0$). Then the restoration heuristic corrects the time slots and produces the correct broadcast trees with significantly lower numbers of conflicts.

4.4.6 The mutation operator

The mutation is primarily intended to bring new genetic material into the population of chromosomes. Typically, very simple mutation operators are used [130]. With an increasing distance of a source-destination pair, the number of possible paths increases exponentially. Therefore, it is necessary to employ a high-quality strategy responsible for testing possible paths and their organization into time slots.

The mutation operator exploits possibilities of some of the proposed heuristics. The mutation operator follows this procedure:

(1) The port-based heuristic is executed. This step prevents conflicts on input and output channels of terminal nodes.

(2) There are two ways to select a mutated gene. The gene is selected with uniform probability in the first half of instances in order to be able to escape from suboptimal solutions. Otherwise, we select only from genes that are causing conflicts in an effort to improve a candidate solution.

(3) One component of the selected gene is selected for mutation.

   (a) Path mutation: a new path between source-destination pair is chosen using Gaussian distribution with the mean value equal to the current path index, and with the variance value equal to one-third of the total number of paths between source-destination pairs. It ensures that paths similar to the current path will be generated with higher probability; on the other hand, any path can still be selected.

   (b) Time slot mutation: first, the transfer swap heuristic is executed. If it does not improve the quality of the solution, the time slot is generated using a uniform random generator.

   (c) For broadcast solutions, there is a possibility to mutate the distributor of the message. The distributor is selected from all terminal nodes participating in the CC using a uniform random generator. Naturally, the restoration heuristic has to be applied before the fitness function is evaluated.

4.5   Optimization tools and parameters adjustments

This section deals with the selection of the most suitable optimization tool for our task. Though the proposed technique is not limited by an optimization technique, we will restrict ourselves to investigate the behaviour of Standard Genetic Algorithm (SGA) [108], Mixed Bayesian Optimization Algorithm (MBOA) [232] and Univariate Marginal Distribution Algorithm (UMDA) [215]. These three tools are the typical representatives of evolutionary and estimation of distribution algorithms. Finally, the most suitable optimization tools will be used in the search for high-quality schedules.

The value of the population size was set to 100 individuals because higher values did not improve the quality of found schedules and did not justify an increased computation time. The binary tournament selects the better half of the current population to form the parent sub-population. The crossover probability was set to 0.9 in the case of SGA and the frequency of the model building to once per generation in cases of MBOA and UMDA. Thereafter, each chromosome is mutated by the proposed mutation operator with probability of 0.9. This operator is responsible for testing and changing possible source-destination paths for particular point-to-point communications. The mutation rate is very high due to a great number of source-destination pairs whose amount growth exponentially with the number of terminal nodes. Finally, the newly generated solutions replace the worse half of the current population.

4.5.1   Experimental comparison of optimization quality

The first experimental work deals with the quality of obtained solutions. The goal of these measurements was the establishment of the most suitable parameters for all selected optimization tools to produce the top quality solutions. Two sufficiently difficult benchmark tasks were chosen; AAS and AAB patterns on the 4D hypercube.

The AAS communication pattern was chosen because of being the most complex variant of M-to-N Scatter communication. Similarly, AAB is the most complex variation of the M-to-N Broadcast communication. The 4D hypercube with 16 nodes was selected due to its symmetry and known optimal solutions for both AAS and AAB communication patterns. The corresponding chromosome lengths for AAS/AAB were 512/786 integer components.

The numbers of communication steps were intentionally set one step below the lower bound for both communication patterns. It makes it possible to compare the algorithms only by the number of conflicts after a finite number of generations. That prevents the algorithm terminating before a given number of generations has been evolved due to discovering the global optima. The graphs show the avarage values of the fitness function (conflicts counts) from all 30 runs after 500,000 generations.

Figure 4.6 compares the achieved solution quality using the best parameters setups for AAS and AAB benchmarks. The best solutions for the AAS benchmark are produced by MBOA. Let us note that solutions produced by UMDA achieve very similar quality. SGA produces slightly worse solutions than both other optimization tools. The best solutions for the AAB benchmark are produced by UMDA. There are larger differences in solution quality in other investigated tools. SGA produces solutions with approximately 30% worse quality than UMDA. The worst solutions are produced by the MBOA algorithm with about twice as worse quality.

For these reasons, UMDA can be declared the best optimization tool for both tested benchmarks.

4.5.2   Experimental comparison of optimization speed

The second experimental work deals with the optimization speed of the proposed optimizations tools. The goal of these measurements was to estimate the time com-

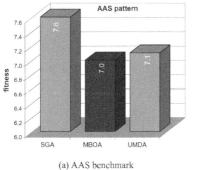
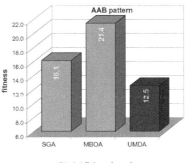

(a) AAS benchmark   (b) AAB benchmark

FIGURE 4.6
The comparison of achieved solution quality obtained by different optimization tools for AAS and AAB benchmarks.

plexity of proposed algorithms and select the fastest one. The same benchmarks as in Section 4.5.1 were used.

The execution times of 500,000 generations of all three optimization tools were measured for both AAB and AAS benchmarks. Then, the measured times were converted into *generation time* and averaged over 30 independent runs. The generation time is the processing time of one generation. In order to be able to compare all three optimization tools, the parameter values have to be set as close as possible with respect to the quality of produced solutions.

Figure 4.7 shows the generation time in milliseconds measured for all three optimization tools. First, we can take note of the incomparable higher generation time of the MBOA tool. MBOA is nearly a hundred times slower than SGA and UMDA. MBOA exploits a very complex probabilistic model to be able to solve hard real-world tasks. It actually achieved the best solution quality for the AAS benchmark. Unfortunately, the creation phase of such a complex model consumes too much time.

Second, SGA is approximately about 30–40% faster than UMDA. Although UMDA belongs to EDA algorithms, its model is very simple. Furthermore, the differences between GA and UMDA are lower for the AAB benchmark.

Finally, AAB generation times are markedly higher than AAS tines. This is caused by a longer genome in the case of AAB (three components per gene against two components per gene in AAS), and also more complex fitness function (restoration heuristic).

For these reasons, the SGA tool can be declared the fastest optimization tool for both tested benchmarks.

4.5.3   Experimental comparison of optimization scalability

This subsection deals with the scalability that indicates the ability to handle growing amount of work in a graceful manner. Two different characteristics of the proposed

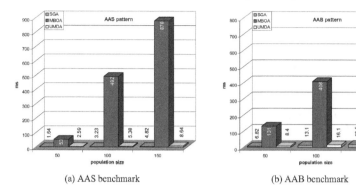

(a) AAS benchmark   (b) AAB benchmark

FIGURE 4.7
The performance of investigated optimization tools with the near optimal parameter setups for AAS and AAB benchmarks.

TABLE 4.2
The success rate SR for various hypercubes for AAS and AAB communication patterns

AAS	Hyper-8	Hyper-16	Hyper-32	Hyper-64
SGA	10%	-	-	-
MBOA	60%	30%	10%	-
UMDA	**100%**	40%	10%	-

AAB	Hyper-8	Hyper-16	Hyper-32	Hyper-64
SGA	70%	20%	-	-
MBOA	**100%**	50%	10%	-
UMDA	**100%**	80%	40%	10%

optimization tools were investigated. We focused on the number of fitness evaluations (NE) needed to get the global optimum and the success rate (SR) representing the percentage of occurrence of the global optimum in ten performed runs. The global optimum is represented by a conflict-free communication schedule (fitness function equals to zero) reaching the lower bound on the number of communication steps.

Both these characteristics were examined using hypercube networks. Five different instances of hypercube topology with 8, 16, 32 and 64 nodes were tested with AAS and AAB benchmarks. Only ten experimental runs were executed because of an excessive time complexity in order of days for some experimental setups.

Figure 4.8 presents the growth of the NE count depending on the hypercube size. The NE characteristics are completed by the SR ones shown in Table 4.2. The configurations of the hypercubes where some tool discovered the global optima in all experimental runs are denoted by crosses. The limits of the proposed tools were found at network size 32-64 nodes in all-to-all communication patterns.

For these reasons, the UMDA tool can be declared as the best scaling tool.

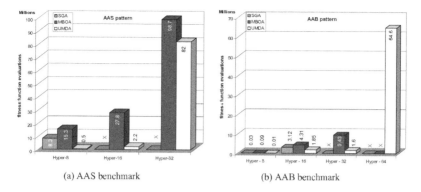

(a) AAS benchmark   (b) AAB benchmark

FIGURE 4.8
The scalability of investigated optimization tools – the numbers of fitness functions evaluations needed to reach global optimum of scatter and broadcast CCs with optimal parameter setups.

4.5.4   Tools assessment

In the search for the most suitable optimization tool we examined in depth the quality of SGA [108], MBOA [232] and UMDA [130].

After a careful comparison of all the proposed tools, we can conclude the UMDA algorithm seems to be the most effective and powerful optimization tool for our task. For this reason, it will be used for all further experiments.

4.6   Experimental results of the quest for high-quality schedules

This section presents the results of the UMDA optimization tool implenting the pro posed techniques when applied to a lot of types of interconnection networks. However, since for many networks studied in this work no analytic methods for scheduling exist, the results can be compared to theoretical lower bounds only. The results are summarized in Table 4.3, Table 4.4 and Table 4.5. Two integers in one cell separated by a slash indicate the lower bound (a smaller integer) has not been reached. A single integer represents identical lower and upper bounds reached by UMDA. Empty cells (denoted by hyphens) are cases that have not been tested yet because of prohibitive time requirements.

4.6.1   Experimental results on common topologies

Table 4.3 presents achieved time complexities in terms of the number of communication steps for a few commonly used topologies such as hypercube, mesh, torus and ring, and some interesting optimal diameter-degree networks such as Petersen [132], Kautz [286], Levi [181] or Heawood [320].

Hypercubes are regular direct topologies with known optimal scheduling, except for OAB reachable by known optimal schedules for any hypercube size. OAS schedules reached the lower bounds in all cases. OAB schedules equaled in the best known algorithms like double tree and Ho-Kao. The optimal schedule reaching 3 steps for a 256-node hypercube is not known up to now. The evolution equalled only the best known suboptimal solution of 4 steps; however the scheduling with 3 steps was almost successful. Only one conflict remained. Unfortunately, it was not possible to remove it even after days of optimization. UMDA also found optimal schedules for AAB up to 64-node and AAS up to 32-node instances. In the case of a 64-node hypercube, the lower bound had to be increase by 3 steps to be able to obtain a conflict-free schedule at least. Let us note the chromosome size for a 64-node hypercube and AAS is 8192 genes with dozens of their possible values!

The bidirectional ring topology, though very simple, is not free from the routing deadlock because the channel dependency graph is not acyclic [82]. This can be seen on a unidirectional as well as a bidirectional ring. The problem can be solved by the introduction of virtual channels [82] and by implementing rules on channel usage. We assume that these rules are adhered to automatically in all our CC schedules and thus the deadlock is avoided. The optimal schedules were found for OAS, OAB and AAB communications. In the case of AAS, optimal solutions were not found, even a suboptimal solution could prevent contentions and in addition improve the performance of a parallel algorithm.

Meshes and tori are also very popular topologies. Meshes can be easily manufactured on a chip due to local interconnections only, but they also have disadvantages. The main one is a lack of node symmetry. Meshes are also irregular networks as the node degree is not constant. While the corner nodes have degree 2, the nodes on the boundary have 3 links and internal nodes 4 links. Therefore, one-to-all schedules for all-port meshes ($k = d$) will need more or less steps accordingly. This is a big difference in comparison to node symmetric tori networks containing wrap-around links. The development of an optimal OAB algorithm in all-port 2D-meshes is difficult because the lower bound is pretty tight. In order to achieve it, we need an algorithm in which every node, once informed, must find in every subsequent step four uninformed nodes and deliver them the message, so that globally all used paths are link-disjoined. Since meshes, unlike tori, are not node-symmetric, there are no elegant algorithms for them. Here again we have to resort to evolutionary optimization that produced successful optimal schedules up to 10×10 mesh and 7×7 tori. If we survey the upper bounds presented in Table 4.3, we will learn that OAS and OAB do not pose a challenge for UMDA. Therefore, the evolution is able to produce novel, maybe up to now unknown, schedules. Finally, the conflict-free AAS schedules are not known even for square tori. AAS schedules were a hard nut to crack, especially in cases of tori. The optimal schedules were designed up to 4×8 (32 node) mesh and only up to 4×3 torus. Rectangular meshes and tori need more relaxed lower bounds, because due to different x and y dimensions not all port can be effectively used.

The evolution has been also applied to several Optimal Diameter-Degree (ODD) topologies that either already found the commercial application (such as scalable Kautz networks) or are potential candidates for example, for NoCs (like non-scalable Petersen ($P = 10$), Heawood ($P = 14$) or Levi ($P = 30$) networks (see Figure 4.9).

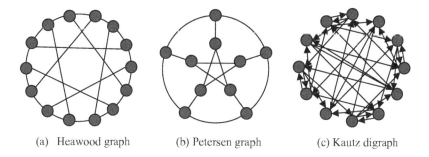

| (a) Heawood graph | (b) Petersen graph | (c) Kautz digraph |

FIGURE 4.9
An example of optimal diameter-degree topologies with 14, 10, and 12 nodes.

To our best knowledge, performance of collective communications on such networks has not been studied so far. The reason may be that until recently [286] these networks have not been used in commercial systems. Consequently, any designed schedule improves the performance of CCs and pushes our knowledge in this area a leap forward.

4.6.2   Experimental results on novel and fat nodes topologies

Table 4.4 summarizes the results obtained on two novel network topologies (Spidergon [149] and AMP [49]) of various size. The table also confronts the evolved schedules on multistage interconnection networks (Omega [170], Butterfly [301], Clos [67] and various types of trees) with the theoretically derived lower bounds. The interesting examples of these networks are shown in Figure 4.10.

Spidergon is also a novel on-chip network architecture suitable for the aggressive on-chip communication demands of SoCs in several application domains and also for networking SoCs. As a ring, it is also not free from deadlock and virtual channels have to be used. From the Table 4.4, it can be seen that UMDA was able to design optimal schedules for OAS and OAB in all instances of the proposed topologies. The obtained schedules for AAB on spidergon also reached the lower bounds. In the case of AAS, UMDA tried to approach the lower bounds as close as possible. In some cases it was verified that the difference between lower and upper (reached) bounds is due to the minimum routing strategy used in our approach. Inclusion of the non-minimum routing would have led to an enormous increase of possible source-destination paths and, therefore, was not explored. However, in some small networks the analysis of the last remaining conflict in fitness function revealed that it can be eliminated if non-minimum routing is used.

A Minimum Path (AMP) configuration is constructed so that the network diam-

TABLE 4.3

The upper bounds $\tau^{CC}(G)$ reached on commonly used topologies (hypercube, ring, mesh, torus) and on special optimal diameter-degree topologies (Petersen, Kautz, Heawood, Levi) using UMDA

Topology	OAS	AAS	OAB	AAB
Hypercube-8	3	4	2	3
Hypercube-16	4	9	2	4
Hypercube-32	7	16	2	7
Hypercube-64	11	35/32	3	11
Hypercube-128	19	-	3	-
Hypercube-256	32	-	4/3	-
Ring-bi-8	4	8	2	4
Ring-bi-16	8	34/32	3	8
Ring-bi-24	12	78/72	4/3	12
Ring-bi-32	16	140/128	4	16
Ring-uni-8	7	28	3	7
Ring-uni-16	15	128/120	4	15
Ring-uni-24	23	300/276	5	23
Ring-uni-32	31	550/496	5	31
2D-Mesh-3x3	4,3,2	6	2,2,2	4
2D-Mesh-3x4	6,4,3	12	3,2,2	6
2D-Mesh-4x4	8,6,4	16	3,2,2	8
2D-Mesh-5x5	12,8,6	32	3,3,3,	12
2D-Mesh-4x8	16,11,8	64	3,3,3	16
2D-Mesh-6x6	18,12,9	56/54	4,3,3	18
2D-Mesh-7x7	24,16,12	-	4,4,3	24
2D-Mesh-8x8	32,22,16	-	4,4,3	32
2D-Mesh-9x9	40,27,20	-	4,4,4	-
2D-Mesh-10x10	50,33,25	-	5,5,5	-
2D-Torus-3x3	2	4	2	3
2D-Torus-3x4	3	6	2	3
2D-Torus-4x4	4	9/8	2	4
2D-Torus-5x5	6	16/15	2	6
2D-Torus-4x8	8	33/32	3	11/8
2D-Torus-6x6	9	30/27	3	9
2D-Torus-7x7	12	-	3	13/12
2D-Torus-8x8	16	-	3	18/16
2D-Torus-9x9	20	-	4/3	-
2D-Torus-10x10	25	-	4/3	-
Petersen-10	3	5	2	3
Kautz-12	4	7	2	4
Heawood-14	5	10/9	2	5
Levi-30	10	31/28	3	10
Kautz-36	12	34/31	3	12
K-Ring-8	2	3	2	2
Midimew-8	2	3	2	2

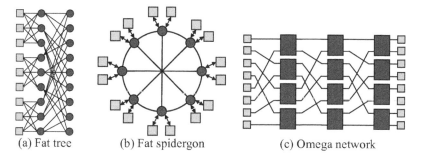

(a) Fat tree   (b) Fat spidergon   (c) Omega network

FIGURE 4.10
An example of indirect networks topologies. CPU cores are denoted by light background while routers are denoted by dark background.

eter and the average inter-node distance is minimized. Let us note that AMP is an asymmetric topology. After checking of AAB results on AMP topology, it could be concluded that UMDA does not offer very good results (e.g., two steps worse schedule for 23 nodes). This problem has been hardly studied. It originates in the asymmetrical nature of AMP where there are some bottlenecks and hot spots that make it impossible to design an optimal schedule. Consequently, the lower bound cannot ever be found. Let us note that this can be a problem of more topologies, not only AMP. The theoretical lower bounds tell us only the lowest possible time complexity of the schedule for a particular CC and topology, but they do not tell us anything about whether it is possible to reach them! On the other hand, UMDA surprisingly designs optimal AAS schedules for all AMP instances.

Multistage Interconnection Networks (MIN) are indirect networks that means the interconnection networks are composed of several processing nodes (processors) and intermediate switches. More formally, MIN is a succession of stages of switching elements (SEs) and interconnection wires connecting P processing (terminal) nodes. SEs in the most general architecture are themselves interconnection networks of small sizes. The interconnection pattern or patterns between MINs stages can be represented mathematically by a set of functions. Examples of such topologies cover an unidirectional Omega [170] (a perfect-shuffle permutation) and Butterfly [301] (butterfly permutation) networks. The main disadvantage of permutation-based MINs is their zero fault-tolerance and high blocking probability. In order to alleviate the bottleneck consisting in only a single path between an input–output pair, the multipath Clos network has been proposed [67]. Here, each network input–output pair can be connected by a path via an arbitrary middle stage. The MINs described so far have unidirectional network links, but bidirectional forms are easily derived as two MINs back-to-back, folded on one another. The overlapping unidirectional links run in different directions, thus forming bidirectional links. A representative of this class is a Fat-tree [178]. Unlike traditional trees in computer science, fat trees resemble real trees because they get thicker near the root.

Obtained schedules for 8-node Omega and Butterfly have met the theoretical lower bound for all classes of collective communications and thus cannot be improved anymore. The limits of simultaneously executable transfers were reached by 12-node and 16-node topologies. For the successful accomplishment of all-to-all communications, UMDA had to add one additional communication step to the theoretically derived value. The Clos network embodies the same problem that has led also in a one step addition. Second, we investigated the ability of the proposed UMDA to discover optimal communication schedules for bidirectional MINs represented by the binary (B-Tree), fat (Fat-Tree), and full binary tree (Full-Tree). Binary trees represent suitable interconnection networks for chip multiprocessor because they need only a very simple link arrangement on a 2D silicon chip. However, as we can see in Table 4.4, their performance rapidly decreases with the number of connected processing nodes. More importantly, the proposed UMDA is able to find optimal communication schedules for most tested binary trees and investigated communication patterns. As can be seen from the table, the lower bounds for the AAB pattern are too tight. Consequently, it was not possible to find optimal solutions. Moreover, there is no certainty that it will be even possible to reach them because the controlled flood does not work for multistage networks.

One way on how to increase the number of interconnected terminal nodes and not make the network more complex is to use a concept of fat topologies. The idea is very simple: all terminal nodes are replaced by multiprocessor nodes, so that more than one CPU core is connected by the shared router to the interconnection network. We can divide 12 CPU cores in spidergon network among 12, 6 or 4 multiprocessor nodes with 1, 2 or 3 CPU per a node, respectively. In this case, only a one-port model is usually assumed. As microelectronic technology is turning to multicore and multi-threaded processors for more performance and less power consumption, networks interconnecting such fat nodes that are of interest.

Table 4.5 summarizes the upper bound attained on a 2-fat one-port hypercube and 2-fat, 3-fat and 4-fat spidergons. The term 2-fat implies two CPUs connected to a switching element. The notation 2-fat-hypercube-8 represents a 3D hypercube with 8 switching elements connected by three input/output links and 16 terminal nodes, two per each switch. It should be noted that neither lower bounds for one-port nor for all-port model apply here. The reason is that we cannot assign 3 network ports (spidergon) of a node explicitly to internal cores. Let us also note that the optimal schedules are not known for the fat spidergon networks so far. Accordingly, the table presents the best know upper bounds of CCs for the fat spidergon networks found by evolution.

4.6.3 Experimental results on fault-tolerant topologies and many-to-many patterns

As the Kautz network is known for its fault tolerance, we have tested performance degradation under a single link fault. A fault diameter of the Kautz-12 with network diameter $D = 2$ is known to be $D + 2 = 4$. This means that among multiple links between any two nodes the longest path is 4. The network performance under a single

TABLE 4.4

The upper bounds $\tau^{CC}(G)$ reached on novel topologies, multistage networks using UMDA

Topology	OAS	AAS	OAB	AAB
Spidergon-6	2	3	2	2
Spidergon-8	3	4	2	3
Spidergon-12	4	9	2	4
Spidergon-16	5	17/16	2	5
Spidergon-20	7	26/25	3	7
Spidergon-24	8	37/36	3	8
Spidergon-28	9	51/49	3	9
Spidergon-32	11	70/64	3	11
Spidergon-36	12	91/81	3	12
Spidergon-64	21	-	3	24
AMP-8	2	3	2	2
AMP-23	6	14	2	8/6
AMP-32	8	22	3	8
AMP-42	11	31	3	14/11
AMP-53	13	46	3	14/13
Omega-8	7	7	3	8
Omega-16	15	16/15	4	16/15
Butterfly-8	7	7	3	7
Butterfly-16	15	16/15	4	16/15
Clos-8	11	12/11	4	12/11
Clos-16	15	16/15	4	16/15
B-Tree-4	3	4	2	3
B-Tree-8	7	16	3	8/7
B-Tree-16	15	64	4	20/15
B-Tree-32	31	256	5	64/31
Fat-Tree-4	3	3	2	3
Fat-Tree-8	7	7	3	7
Fat-Tree-16	15	15	4	15
Fat-Tree-32	31	32/31	5	31
Full-Tree-7	6,4,3	12	3,2,2	7
Full-Tree-15	14,12,8,7	56	3,3,3,3	15
Full-Tree-31	30,28,24,16,15	240	4,4,4,4,4	31
Full-Tree-63	62,60,56,50,48,32	992	5,5,5,5,5,5	64

TABLE 4.5

The upper bounds $\tau^{CC}(G)$ reached on fat topologies
using UMDA

Topology	OAS	AAS	OAB	AAB
2-fat-hypercube-4	7	8	3	7
2-fat-hypercube-8	15	17	4	15
2-fat-hypercube-16	31	36	6	33
2-fat-hypercube-32	63	-	7	-
2-fat-spidergon-6	11	12	4	12
2-fat-spidergon-8	15	17	4	16
2-fat-spidergon-10	19	25	5	20
2-fat-spidergon-12	23	37	5	24
2-fat-spidergon-14	27	50	5	29
2-fat-spidergon-16	31	66	6	37
2-fat-spidergon-18	35	86	6	42
3-fat-spidergon-4	11	11	4	11
3-fat-spidergon-8	23	38	5	24
3-fat-spidergon-12	35	82	6	42
4-fat-spidergon-4	15	16	4	16
4-fat-spidergon-6	23	38	5	24
4-fat-spidergon-8	31	64	6	34

link fault is given in Table 4.6 (with node 0 as the source node for OAB and OAS), but the network could operate even under a double link fault. In any case, when the link fault is detected, the new schedule could be computed in 20 seconds on a single processor and then the cluster could continue with a lower performance (Table 4.8).

Besides one-to-all and all-to-all communication patterns, we can also find a lot of communication patterns that engage only a subset of nodes. Any CC executed on an indirect (fat, multistage) network can be actually seen as M-to-N communication, because switching nodes are excluded from the communication. Any combination of all-port and one-port nodes is allowed by our technique. We investigated only a couple of possible M-to-N arrangements, but it can be simply demonstrated that the proposed technique is able to design schedules for M-to-N communication with any distribution of transmitters and receivers. The topologies of interest were the 8-node all-port hypercube and 8-node all-port spidergon. The upper bounds of obtained schedules are summarized in Table 4.7.

Table 4.8 shows average execution times of the UMDA during five successful runs on several networks. All experiments were realized in sequential manner on IBM Blade servers equipped with 2x dual core AMD Opteron 275 processors and supplied by 4 GB DDR2 RAM at 800 MHz. OAS communication is relatively easy; a solution always takes less than one second. For OAB communication, the values are less than one second for simple network topologies. The longest execution time (hypercube-32) is about 41 seconds. On the other hand, a suitable solution for all-to-all communication takes a much longer time, especially for the AAS communication. An exponential increase of the execution time with the network size can be observed.

TABLE 4.6

Performance of the Kauz-12 network
with a single faulty link (reduced
performance is in bold). The upper
bounds obtained by UMDA

link	OAS	AAS	OAB	AAB
no fault	4	7	2	4
0→3	**6**	**9**	**3**	**6**
0→5	**6**	**9**	**3**	**6**
0→4	**6**	**9**	**3**	**6**
1→7	4	**9**	2	**6**
1→6	4	**9**	2	**6**
1→8	4	**9**	2	**6**
2→B	4	**9**	2	**6**
2→A	4	**9**	2	**6**
2→9	4	**9**	2	**6**
3→0	4	**9**	2	**6**
3→1	**5**	**9**	2	**6**
3→2	**5**	**9**	2	**6**
5→7	4	**9**	2	**6**
all other	4	**9**	2	**6**

TABLE 4.7

The upper bounds $\tau^{CC}(G)$ for the 8-node hypercube and
spidergon for a few M-to-N CC patterns

Topology	Pattern	MMS	MMB
	Within the same base	2	2
Hypercube-8	Between the bases	4	2
	Lower base to all	4	3
	Diagonal to the lower base	2	2
	Within the left half	2	2
Spidergon-8	Left half to right half	3	3
	Left half to all	3	3
	Odd nodes to even nodes	3	2

TABLE 4.8

Execution times of UMDA in seconds, minutes, hours and days on AMD Opteron 275 with 4 GB RAM (average values during 5 successful runs)

Topology	OAS	AAS	OAB	AAB
Ring-8	<1s	5m6s	<1s	<1s
Uni-Ring-8	<1s	57s	<1s	<1s
Spidergon-8	<1s	2s	<1s	<1s
Petersen-10	<1s	12s	<1s	2s
Kautz-12	<1s	23s	<1s	3s
Heawood-14	<1s	9m17s	<1s	5s
Spidergon-16	<1s	22m36s	3s	1m41s
Levi-30	<1s	1d6h	3s	2h2m
Hypercube-32	<1s	4d5h	41s	28m38s
Kautz-36	<1s	3d5h	20s	9h50m

For the most complex topologies it can easily reach impractical values of (days). The execution times are the main restriction of the proposed evolutionary approach of CC scheduling and puts a limit on the size of solvable networks.

An example of optimal AAB and AAS communication schedules on 8-node AMP is shown in Figure 4.11 and Figure 4.12. The AMP topology is a result of genetic graph optimization [49]. A minimum path (AMP) [50] configuration is constructed so that the network diameter and the average inter-node distance is minimized. This principle is maintained even at the expense of the loss of regularity in the system. The AMP networks have been found for node count $P = 5, 8, 12, 13, 14, 32, 36, 40, 53, 64, 128$ and 256. The SC node denotes a system controller (host computer) that sends input data to processing nodes and collects results. Each processing node can communicate simultaneously on four bi-directional full duplex links.

In the first communication step, all nodes inform some of their neighbors. In the second step, the messages are propagated further through the network to other nodes not yet informed. An example of optimal AAS communication schedule is shown in Figure 4.12. The optimal schedule can be executed in three communication steps.

4.7   Conclusions

The importance of point-to-point interconnection networks steadily grows. These networks are directly predestined to force out bus-based interconnections from all levels of system design including Systems on Chip (SoC) due to their simple manufacturability, reliability, fault-tolerance, low latency and high throughput. Many real multiprocessor systems have been based on NoC (e.g., Tilera Tile, Sicortex SC systems, IBM Cell processor). Interconnection networks have also been used at chip

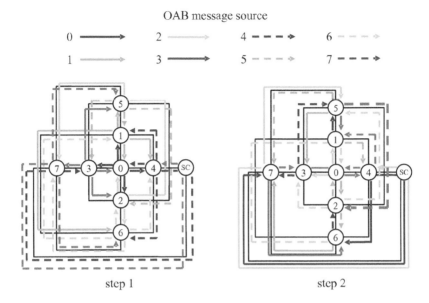

FIGURE 4.11
An example of optimal AAB schedule on the 8-nodes AMP topology.

interconnection level (e.g., Intel QuickPath or AMD HyperTransport). The transitions from bus-based systems can be also seen in module interconnections (e.g., PCI Express).

One of the key issues in this area is the performance of communications among nodes. Since each parallel task requires communication and synchronization mechanisms, the communication latency should be minimized while the network throughput should be maximized. Most communication and synchronization mechanisms can be implemented as collective communications (CC) based on scatter/gather and broadcast/reduce services. For example, the data distribution among nodes can be implemented by one-to-all scatter, the parallel algorithm control by one-to-all broadcast, results collection by all-to-one gather or reduce, barrier synchronization by all-to-all gather followed by one-to-all broadcast or all-to-all broadcast, etc.

The time complexity of a particular collective communication pattern can be estimated from the network parameters. The lower bound on time complexity can be expressed in terms of the number of synchronized communication steps (time slots). Although the lower bound can be obtained relatively easily, the design of such a communication schedule that meets the lower bound is still an outstanding problem because the schedule has to prevent link blocking, deadlock and livelock [140].

Although many approaches that implement (near) optimal schedules of CC have been published to date, the authors have concentrated either on only a certain topology, such as hypercube [126] or mesh, or have not been able to work with irregular

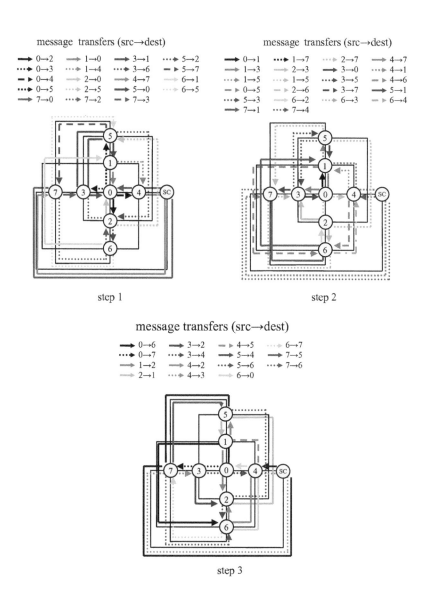

FIGURE 4.12
An example of optimal AAS schedule on the 8-nodes AMP topology.

topologies, many-to-many communication patterns, the all-port model or indeterministic routing. A universal approach has not existed until now.

Therefore, the hypothesis of the chapter has been formulated. Evolutionary design is able to produce optimal or near optimal communication schedules comparable or even better than those obtained by a conventional design for the networks sizes of interest. Moreover, evolutionary design reduces many drawbacks of present techniques and invents still unknown schedules for an arbitrary topology and scatter/broadcast communication patterns [143].

Because the evolutionary algorithms are highly sensitive to a suitable parameters setup, a comprehensive study has been accomplished in Section 4.5. The results have shown that the most suitable one is the UMDA algorithm [215]. The results have also shown that the evolution is controlled mainly by a mutation operator and that the sufficient population size is relatively small.

Chapter 4.6 has presented the achievements of the UMDA algorithm. The postulated hypothesis has been completely proved in this chapter. UMDA has reinvented many important optimal scheduling techniques like the recursive doubling [198] on hypercubes, rings, tori or TAD Tree [300] approach for tori and hypercubes. It also equaled the Ho-Kao algorithm [126] and ST algorithm [230]. Moreover, many unknown optimal schedules have been invented namely for optimal diameter-degree networks, irregular meshes, fat topologies, indirect topologies, and many-to-many communication pattern. If it was not possible to design an optimal schedule, a suboptimal one was found in reasonable time. More results and comments can be found in a recently published PhD thesis [143].

Since the schedules are designed for the source-based routing technique, the wormhole switches can remain without any change. The schedules can be uploaded into processing nodes that can exploit them to perform the communications as fast as possible. The designed schedules may also serve for writing high-performance communication functions for a concrete topology. Consequently, this function can be included into, for example, a well-known OpenMPI [302] library. Otherwise, when a promising topology would be introduced, the proposed technique could be simply applied to this topology and accelerate the communication on it.

4.7.1   Contributions of the proposed technique

This section summarizes all notable contribution of the proposed technique [143].

(1) The new method for scheduling of collective communication that prevents link blocking, deadlock and livelock and is not limited by a topology or a distribution of cooperating nodes has been proposed.

(2) Requirements on components of a tool for optimizing CC schedules have been formulated. Hence a lot of optimization tools based on classical artificial intelligence or heuristic methods like hill climbing as well as stochastic methods like simulating annealing, particle swarm optimization, and artificial ant colonies can be exploited.

(3) The integrity of the technique is given by its capabilities to design optimal or at least sub-optimal schedules for many kinds of network topologies and communication patterns that have never been solved so far. The technique can be particularly applied to:

   (a) Regular topologies (e.g., hypercube and torus)

   (b) Irregular topologies (e.g., mesh)

   (c) Multi-stage topologies (e.g., Omega, Clos)

   (d) Topologies with fat nodes of any type (e.g., fat tree, fat hypercube)

   (e) One-port, all-port, and even k-port model (by an additional limitation of port based heuristic) are acceptable

   (f) Minimal and even non-minimal routing (only for small scale topologies)

   (g) Simplex, full duplex links and even half-duplex (by a simple modification of conflict counting function that would both make difference between channel directions)

   (h) Any distribution of cooperating nodes on the topology (one-to-all, many-to-many, all-to-all)

   (i) Faulty nodes and faulty links (a new operational state can be restored after a link or a node fault)

   (j) And many others

(4) Many optimal communication schedules have been re-invented with the help of the technique based on the UMDA algorithm. Let us recall the recursive doubling algorithm or time-arc disjoint broadcast trees at least. Very significant success has been in reproducing results of the Ho-Kao algorithm for OAB hypercube [126].

(5) Many still unknown optimal schedules have been designed by the technique (e.g., rectangular tori, meshes, spidergons, AMP, optimal diameter-degree networks).

4.7.2   Future work

Although the proposed technique covers a large area of the scheduling of collective communication on wormhole networks, there are naturally many other possible extensions of the technique.

   An introduction of the channel capacity (throughput) is one of many possible future extensions. Only the channels with the same capacity have been presumed in this work, but in real-world applications the channels can be organized into more levels of different capacity (e.g., backbone links, inter-cluster links, local interconnection). Accordingly, the capacity of the channels should be taken into account in the fitness function to allow more transfers at a time with respect to the channel capacity.

Only scatter or broadcast CCs have been investigated in this work because they are most frequently used. A few additional communication services would be useful in real-world applications. Let us note for example permutation and scan communications or application specific CCs created by a collection of point-to-point communications. The input file structure would have to be modified to be able to handle a general CC. The source destination pairs of all message transfers would have to be provided. A corresponding chromosome would be based on many-to-many scatter one.

For some networks, it could also be beneficial to use the combining model that was not used in this work in order to keep the switches as simple as possible. Message combining reduces the total number of messages, making each node send fewer messages of a larger size. Reducing the number of steps can improve communication performance in the case of short messages when the start-up delays dominate in CC times. Unfortunately, the combination of messages completely changes the principles of the scheduling and a new scheduling technique would have to be designed.

Another direction for future research could be targeted to an improvement of acceleration and restoration heuristics. Limitations of the technique have been observed mainly at AAS communications, so additional research in this area could be beneficial. Improved results could also be obtained by employing a more sophisticated optimization tool, such as particle swarm optimization (PSO).

Acknowledgment

This research has been partially supported by the research grants "Safety and security of networked embedded systems applications," GA 102/08/1429 of Czech Science Foundation (2008-10), "Natural Computing on Unconventional Platforms," GP103/10/1517, Czech Science Foundation (2010-13), "Secured, reliable and adaptive computer systems," BUT FIT grant FIT-10-S-1 (2010) and the research plan "Security-oriented research in information technology," MSM 0021630528 (2007-13).

5

Formal Aspects of Parallel Processing on Bio-Inspired on-Chip Networks

P. C. Vinh

NTT University, Vietnam

CONTENTS

A networking-on-chip paradigm is currently at crucial point in its evolution: *autonomic networking-on-chip* (ANoC), marked by the increasing developments of *bio-inspired on-chip networks* (shortly called as BioChipNets). In ANoC, every computation is split up into tasks that run simultaneously on multiple *IP* (intellectual property) *cores* communicating over a BioChipNet. Hence, ANoC is a form of parallel processing in decentralization, nondeterminism and dynamicity of the network. The overarching goal of ANoC is to support such BioChipNets capable of management and high performance. Meeting this grand challenge of ANoC requires a fundamental approach to the aspects of tasks and data parallel not tackled before. To this end, taking advantage of the categorical structures we establish, in this chapter, a firm formal basis for specifying tasks parallel, data parallel, core-to-core structures and self-organization on BioChipNets. All of these are to formalize parallel processing on BioChipNets.

5.1   Introduction

Communicating based on bio-inspired on-chip networks (BioChipNets) is currently on spot and imply an increased decentralization in managing the BioChipNet behavior [194]. *Autonomic networking-on-chip* (ANoC) is an essential networking-on-chip paradigm to keep such networks capable of management and high performance [83]. In fact, the problem is that many BioChipNets make central or global control impossible. For example, the information needed to make decisions cannot be gathered centrally to support tasks that run simultaneously on multiple IP cores in a BioChipNet infrastructure [233]. In such networks, ANoC is only possible when computational agents autonomously interact and coordinate with each other to maintain properly the required computations [142, 195, 311, 313]. The essence of ANoC is to enable the computational agents to execute parallel tasks and deliver parallel resources while interacting and coordinating with each other.

Hence, for BioChipNets, one of major challenges is how to support tasks parallel and data parallel in the face of changing computation needs and objectives. In other words, how can a BioChipNet perform parallel tasks and process parallel data in accordance with context-awareness and goal-driven computational mechanisms?

Dealing with this grand challenge of BioChipNets requires a well-founded modeling and in-depth analysis on the paradigm of ANoC. With this aim, we develop a firm formal approach in which the notions of *task*, *data type*, *agent* and *self-configuration* are formulated and specified in categorical structures known as a categorical approach of tasks and data parallel processing.

In ANoC, every computation is divided into tasks that run simultaneously on multiple IP cores communicating over a BioChipNet. Hence, ANoC is a form of parallel processing, but traditional parallel processing is most commonly used to describe tasks running simultaneously on multiple processors in the same chip [104, 142]. Moreover, major distinctions stemming from distributed environments are that tasks often must deal with *decentralization, nondeterminism and dynamicity* [73, 104, 142] on the network.

This chapter breaks new ground in dealing with the *core-to-core* and *agent-based networking techniques* [311] for ANoC [104, 142, 193] using categorical approach of tasks and data parallel processing – the firm formal method applicable to a wide variety of BioChipNets.

The major contribution of the chapter is to propose some applied categorical structures of tasks parallel and data parallel for parallel processing on BioChipNets which, to the best of our knowledge, have never been tackled thoroughly in this emerging field. In fact, by this approach, category theory is applied to deal in an abstract way with algebraic objects and relationships between them for specifying tasks parallel and data parallel on BioChipNets. It turns out that, for instance, a theory of self-organization investigated by E. Anceaume et al. in [6] or formal model of autonomous agent systems presented by Y. Wang in [319] are just algebraic objects of category. Hence, for specifying, analyzing and verifying tasks and data parallelism, the categorical approach becomes much better-approaching than other ones in theory of algebras.

From an applicative aspect of the approach, moreover, formalizing tasks parallel and data parallel on BioChipNets is a categorical specification of middleware that can be used to develop implementations for core-to-core and agent-based networking. The formalization describes what the middleware system should do, not how such middleware system should do it. For core-to-core or agent-based networking, a middleware system design cannot ever be developed in isolation, but only with respect to a given specification. Therefore, this formalization is applicable to the following.

Given a specification, scalable verification techniques (such as model checking, equivalence checking, satisfiability checking, theorem proving, constraints solving, slicing and distributing the state space, and so on) can be used to justify that a candidate middleware system design is correct with respect to the specification. This has the benefit that incorrect candidate middleware system designs can be revised before any further endeavor is made to implement the design actually.

Another application is to use provably stepwise refinement to transform, in verifiable stages, a specification into a middleware system design, and finally into an actual implementation, that is correct by construction.

Apparently, it is able to validate a specification by proving theorems about properties that the specification is expected to expose. If correct, these theorems increase our knowledge over the specification and its relationship with the underlying problem domain of BioChipNets. Otherwise, the specification needs to be adapted to reflect better our insight over BioChipNets including implementing the specification.

Furthermore, there are the mutual applications of ANoC and formal verification. On the one hand, ANoC provides suitable abstractions and platforms to support the construction of scalable verification tools. On the other hand, the scalable verification techniques support the design and analysis of the communication protocols needed to reach more dependable robust and predictable behavior of BioChipNets.

5.2 Outline

The chapter is a reference material for readers who already have a basic understanding of BioChipNets and are now ready to know the novel approach for formalizing parallel processing on BioChipNets using categorical language.

Formalization is presented in a straightforward fashion by discussing in detail the necessary components and briefly touching on the more advanced components. Several notes explaining how to use the formal aspects, including justifications needed in order to achieve the particular results, are presented.

We attempt to make the presentation as self-contained as possible, although familiarity with the notion of tasks and data parallel on BioChipNets is assumed. Acquaintance with the algebra and the associated notion of categorical language is useful for recognizing the results, but is almost everywhere not strictly necessary.

The rest of this chapter is organized as follows: In Section 5.4, we recall some concepts from the category theory used in the chapter. Section 5.3 includes some major work related directly to the content of the chapter. Section 5.5 presents parallel composition of tasks on BioChipNets, category of BioChipNet tasks together with the notion of diagram chasing, then category of core-to-core networks together with categorical aspects of self-configuration. In Section 5.6, we present type of task, type based-agent and category of data types. Then extensional monoidal structure and symmetry monoidal structure of the category of data types are constructed in order to consider the significant properties of data parallel on BioChipNets. In Section 5.7, we briefly discuss a direction of further developments in the future. Finally, a short summary is given in Section 5.8.

5.3 Related work

Systems on a chip (SoC) are complex embedded systems focusing less on computation and increasingly on communication. This shifts design style based on platforms [153] (i.e. design templates) to communication-based design [73, 231, 273]. In this new paradigm, NoC has emerged as a new type of interconnect that can solve this problem [72, 104, 142, 201]. Importantly, formal methods for NoC attract rising attention from community of NoC researchers [44, 110, 111, 125, 236, 264, 268, 269].

In fact, in [110], Æthereal NoC is introduced as an example to identify when and where formal methods can play a role in the field of NoC research. A functional approach to the formal specification of NoCs is proposed in [268]. An NoC includes several levels of the protocol stack such as physical layer, data link layer, network layer and transport layer. When the NoC protocol is composed of such different aspects then guaranties about lossless delivery, deadlock freeness and other Quality of Service properties are required. Hence it follows that formal verification is concerned [244, 264, 309]. Most proposed verification solutions for performance, traffic or behavior analysis for NoCs are simulation or emulation-oriented [111,236]. A formal verification of NoCs by means of a mechanized proof tool, the ACL2 theorem prover, is developped [44, 269]. A methodology using theorem proving techniques for the formal verification of NoC communications with non-minimal adaptive routing is proposed in [125].

In parallel processing, communicating processes are regularly classified into two main paradigms: *accessing shared variables* or *passing messages* [325]. While this chapter is built on a very large achievement of existing work, it mainly stems from the following three sources.

Among results of parallel processing with shared variables, first we take account of Xu and He's work [325] on laws of parallel programming with shared variables which in turn was influenced by C.A.R. Hoare et al. [128] in laws of programming.

In parallelism of passing messages, second we are interested in Hoare's CSP [127], Milner's work [203] on calculus of concurrent programming, and Baeten and Weijland's work [16] on Algebra of Communicating Processes (ACP) among others.

Finally, the choice of the underlying formalization requires a close look at models for parallel processing. Hence, our interest centers on Michel's approach to distributed systems taking advantage of category theory [200]. In fact, categories were first described by Samuel Eilenberg and Saunders Mac Lane in 1945 [171] but have since grown substantially to become a branch of modern mathematics. Category theory spreads its influence over the development of both mathematics and theoretical computer science. The categorical structures themselves are still the subject of active research, including work to increase their range of practical applicability.

5.4 Basic concepts

In this section, we recall some concepts from the category theory [3, 9, 31, 171, 182] used in this chapter.

5.4.1 Category definition

5.4.1.1 Category as a graph

A category $\mathbf{C}$ can be viewed as a graph $(Obj(\mathbf{C}), Arc(\mathbf{C}), s, t)$, where

- $Obj(\mathbf{C})$ is the set of nodes we call *objects*.

- $Arc(\mathbf{C})$ is the set of edges we call *morphisms*.

- $s, t : Arc(\mathbf{C}) \longrightarrow Obj(\mathbf{C})$ are two maps called *source* (or *domain*) and *target* (or *codomain*), respectively.

We write $f : X \longrightarrow Y$ when f is in $Arc(\mathbf{C})$ and $s(f) = X$ and $t(f) = Y$.

Explanation on terminology: An object in the category is an algebraic structure such as a set. We are probably familiar with some notations for finite sets: {*Student A, Student B, Student C*} is a name for the set whose three elements are *Student A, Student B, Student C*. Note that the order in which the elements are listed is irrelevant.

A morphism f in the category consists of three things: a set X, called the source of the morphism; a set $\mathcal{Y}$, called the target of the morphism and a rule assigning to each element x in the source an element y in the target. This y is denoted by $f(x)$, read "f of x." Note that the morphism is also called the *map, function, transformation, operator* or *arrow*. For example, let $X = $ {*Student A, Student B, Student C*}, $\mathcal{Y} = $ {*Math, Physics, Chemistry, History*} and let f assign each student his or her favorite subject. The following internal diagram is an illustration.

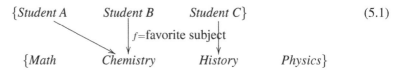

$$\begin{array}{llll} \{Student\ A & Student\ B & Student\ C\} & \quad(5.1) \\ & f\text{=favorite subject} & \\ \{Math & Chemistry & History & Physics\} \end{array}$$

This states that the favorite subject of the *Student C* is *History*, written by f(*Student C*) = *History*, while *Student A* and *Student B* prefer *Chemistry*. There are some important properties of any morphism

- From each element in the source {*Student A, Student B, Student C*}, there is exactly one arrow leaving.

- To an element in the target {*Math, Physics, Chemistry, History*}, there may be zero, one or more arrows arriving.

It is possible that the source and target of the morphism could be the same set. The following internal diagram is an example.

$$\begin{array}{lll} \{Student\ A & Student\ B & Student\ C\} & \quad(5.2) \\ & & e\text{=favorite classmate} \\ \{Student\ A & Student\ B & Student\ C\} \end{array}$$

and, in the case, the morphism is called an *endomorphism* whose representation is available as in

$$\{Student\ A \longrightarrow Student\ B \longleftarrow Student\ C\} \quad\quad (5.3)$$

5.4.1.2 Identity morphism and composition of morphisms

Associated with each object X in $Obj(\mathbf{C})$, there is a morphism $1_X = X \longrightarrow X$, called the *identity* morphism on X, and to each pair of morphisms $f : X \longrightarrow \mathcal{Y}$ and $g : \mathcal{Y} \longrightarrow \mathcal{Z}$, there is an associated morphism $f;g : X \longrightarrow \mathcal{Z}$, called the *composition* of f with g. The representations in (5.4) include the external diagrams of identity morphism and composition of morphisms.

$$1_X$$

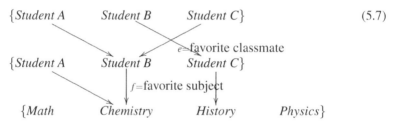

$$X \xrightarrow{f} Y \xrightarrow{g} Z \qquad (5.4)$$
$$f;g$$

Explanation on terminology: Here are the corresponding internal diagrams of the identity morphism.

$$\{Student\ A \qquad Student\ B \qquad Student\ C\} \qquad (5.5)$$

$$\{Student\ A \qquad Student\ B \qquad Student\ C\}$$

Or

$$\{Student\ A \qquad Student\ B \qquad Student\ C\} \qquad (5.6)$$

And here, the composition of morphisms is described in the internal diagram

$$\{Student\ A \qquad Student\ B \qquad Student\ C\} \qquad (5.7)$$

$e=$favorite classmate

$$\{Student\ A \qquad Student\ B \qquad Student\ C\}$$

$f=$favorite subject

$$\{Math \qquad Chemistry \qquad History \qquad Physics\}$$

Or, in the external diagram $X \xrightarrow{e} X \xrightarrow{f} Y$. From diagram (5.7), we can obtain answers for the question "What should each student support to his or her favorite classmate for subject?" In fact, the answers are such as " *Student A* likes *Student B*, *Student B* likes *Chemistry*, so *Student A* should support *Chemistry*," "*Student B* likes *Student C*, *Student C* likes *History*, so *Student B* should support *History*" and "*Student C* likes *Student B*, *Student B* likes *Chemistry*, so *Student C* should support *Chemistry*."

The composition of two morphisms e and f means that e and f are combined to obtain a third morphism $X \xrightarrow{e;f} Y$. This is represented in the following internal diagram.

$$\{Student\ A \qquad Student\ B \qquad Student\ C\} \qquad (5.8)$$

$e;f$

$$\{Math \qquad Chemistry \qquad History \qquad Physics\}$$

where, for example, $e;f(Student\ B) = History$ is read as "the favorite subject of the favorite classmate of *Student B* is *History*."

5.4.1.3 Identity and associativity for composition of morphisms

The following equation must hold for all objects X, Y in $Obj(\mathbf{C})$ and morphism $f : X \longrightarrow Y$ in $Arc(\mathbf{C})$:

$$\text{Identity:} \qquad 1_X; f = f = f; 1_Y \qquad (5.9)$$

$$1_X \, \underset{}{\circlearrowleft} X \xrightarrow{\ f\ } Y \; = \; X \xrightarrow{\ f\ } Y \; = \; X \xrightarrow{\ f\ } Y \, \circlearrowright \, 1_Y$$

The following equation must hold for all objects X, Y and Z in $Obj(\mathbf{C})$ and morphisms $f : X \longrightarrow Y$, $g : Y \longrightarrow Z$ and $h : Z \longrightarrow T$ in $Arc(\mathbf{C})$:

$$\text{Associativity:} \qquad (f; g); h = f; (g; h) \qquad (5.10)$$

$$X \xrightarrow{\ f\ } Y \xrightarrow{\ g\ } Z \xrightarrow{\ h\ } T \quad = \quad X \xrightarrow{\ f\ } Y \xrightarrow{\ g\ } Z \xrightarrow{\ h\ } T$$
$$\underbrace{\qquad\quad}_{f; g} \qquad\qquad\qquad\qquad \underbrace{\qquad\quad}_{g; h}$$

5.4.2 Functor

Functor is a special type of mapping between categories. Functor from a category to itself is called an *endofunctor*. Note that, in this chapter, when the notion of endofunctor dominates throughout in use, then we can name them as the functor, for short, without any confusion. The functors are also viewed as morphisms in a category, whose objects are smaller categories.

A *multifunctor* is a generalization of the functor concept to n arguments. Specially, a *bifunctor* is a multifunctor with $n = 2$.

There are two kinds of functors distinguished by the way they treat morphisms to be *covariant* and *contravariant*. A functor T is covariant if for each source morphism $X \xrightarrow{\ f\ } Y$ the target morphism has the form $\mathsf{T}\, X \xrightarrow{\ \mathsf{T} f\ } \mathsf{T}\, Y$. A functor T is contravariant if for each source morphism $X \xrightarrow{\ f\ } Y$ the target morphism has the form $\mathsf{T}\, X \xleftarrow{\ \mathsf{T} f\ } \mathsf{T}\, Y$.

5.4.3 Isomorphism

A morphism $f : X \longrightarrow Y$ in the category $\mathbf{C}$ is an *isomorphism* if there exists a morphism $g : Y \longrightarrow X$ in that category such that $f; g = 1_X$ and $g; f = 1_Y$.

$$X \xrightarrow{\ f\ } Y \xrightarrow{\ g\ } X \quad \text{and} \quad Y \xrightarrow{\ g\ } X \xrightarrow{\ f\ } Y \qquad (5.11)$$
$$\underbrace{\qquad\qquad}_{f; g = 1_X} \qquad\qquad\qquad \underbrace{\qquad\qquad}_{g; f = 1_Y}$$

That is, if the following diagram commutes.

$$1_X \,\big(\!\!\bigcirc X \underset{f}{\overset{g}{\rightleftarrows}} Y \bigcirc\!\!\big)\, 1_Y \tag{5.12}$$

5.4.4   Natural isomorphism

Let **C** and **C′** be two categories. Consider a parallel pair

$$\mathbf{C} \underset{T'}{\overset{T}{\rightrightarrows}} \mathbf{C'} \tag{5.13}$$

of functors of the same variance. Two functors of T and T′ are *naturally equivalent* (also called *naturally isomorphic*) if there is an inverse pair of natural isomorphisms between them. In other words, the inverse pair of natural isomorphisms between T and T′ is a pair

$$T \underset{\zeta_-}{\overset{\eta_-}{\rightleftarrows}} T' \tag{5.14}$$

such that for each object X in **C** the morphisms

$$T\,X \underset{\zeta_X}{\overset{\eta_X}{\rightleftarrows}} T'\,X \tag{5.15}$$

are an inverse pair of isomorphisms in **C′**.

5.4.5   Element of a set

For any set A, $x \in A$ iff $1 \overset{x}{\longrightarrow} A$ (or $x : 1 \longrightarrow A$) where 1 denotes a singleton set. Focus on one element of {*Math, Physics, Chemistry, History*}, say {*subject*}, and call this set "1." Let us see what the morphisms from 1 to {*Math, Physics, Chemistry, History*} are. There are exactly four of them.

$$\{subject\} \quad \{Math \quad Chemistry \quad History \quad Physics\} \tag{5.16}$$
$$\underbrace{\qquad\qquad}_{Math}\!\nearrow$$

$$\{subject\} \quad \{Math \quad Chemistry \quad History \quad Physics\} \tag{5.17}$$
$$\underbrace{\qquad\qquad\qquad\qquad}_{Chemistry}\!\nearrow$$

$$\{subject\} \quad \{Math \quad Chemistry \quad History \quad Physics\} \quad (5.18)$$

$$History$$

$$\{subject\} \quad \{Math \quad Chemistry \quad History \quad Physics\} \quad (5.19)$$

$$Physics$$

By this way, we can write $1 \xrightarrow{0} \mathbb{N}_0$ (or $0 : 1 \longrightarrow \mathbb{N}_0$) for $0 \in \mathbb{N}_0$, $1 \xrightarrow{i} \mathbb{N}$ (or $i : 1 \longrightarrow \mathbb{N}$) for $i \in \mathbb{N}$ and so on.

5.5 Processing BioChipNet tasks

We present this section concentrating on parallel composition of tasks on BioChip-Nets, on category of BioChipNet tasks together with the notion of diagram chasing, then on category of core-to-core networks together with categorical aspects of self-configuration.

5.5.1 Parallel composition of BioChipNet tasks

Autonomic networking-on-chip is a type of parallel processing. Both forms of processing require dividing a computation into tasks that can run simultaneously, but autonomic computations on BioChipNets often must deal with challenges as distributed environments, decentralization, nondeterminism and dynamicity on BioChipNets, whereas traditional parallel processing is most commonly used to describe computation tasks running simultaneously on multiple processors in the same chip.

Definition 1 *A parallel composition of tasks on some BioChipNet is a binary operation (denoted as $\_ \parallel \_$) between two tasks on the BioChipNet which groups them together as being parallel in the distributed environments. Hence, let a and b be arbitrary tasks on the BioChipNet, then $a \parallel b$ denotes that a is parallel to b.*

It follows that the parallel composition of a series of tasks $a_1, a_2, \ldots, a_n$, $a_1 \parallel (a_2 \parallel (\ldots \parallel a_n) \ldots)$, is simply written as $a_1 \parallel a_2 \parallel \ldots \parallel a_n$ or $\parallel_{1 \leqslant i \leqslant n} a_i$. Sometimes, the notation of $\parallel_0$ is used to denote a special task *skip* that has no effect on any state of BioChipNet, and terminates immediately.

The algebraic laws [128, 325] governing the behavior of $a \parallel b$ are exceptionally simple and regular. The following are three laws related to our development in this section.

- *Law on symmetry*: This law expresses the logical *symmetry* between two tasks

$$a \parallel b = b \parallel a \qquad (5.20)$$

• *Law on associativity*: This law shows that when three tasks are assembled, it does not matter in which order they are put together

$$(a \parallel b) \parallel c = a \parallel (b \parallel c) \tag{5.21}$$

• *Law on identity*: Composition with *skip* makes no difference

$$a \parallel skip = a \tag{5.22}$$

5.5.2   BioChipNet Tasks

For every task a and b on a BioChipNet $\mathcal{N}$, if a is parallel to b then we define a labeled arrow from a to b as $a \xrightarrow{a\parallel b} b$ in which its label is $a \parallel b$. Category of tasks of $\mathcal{N}$ is founded upon the abstraction of the labeled arrow called *morphism*. Here, the labeled arrow "$\xrightarrow{a\parallel b}$" is the morphism between a and b. We usually use two following diagrams to specify this morphism.

$$a \xrightarrow{a\parallel b} b \qquad \text{or} \qquad a \parallel b : a \longrightarrow b \tag{5.23}$$

where source and target of the morphism $a \parallel b$ are the tasks a and b, respectively. Such directional structures occur widely in representation of this chapter.

By describing structures in terms of the existence and characteristics of morphisms _ $\parallel$ _, categorical structures of BioChipNet tasks achieve their wide applicability. The usual method of mathematical description is by reference to the internal structure of BioChipNet tasks. The applicability of this description is then limited to BioChipNet tasks supporting such structure. Categorical descriptions make no assumption about the internal structure of BioChipNet tasks, but they purely ensure that whatever structure of BioChipNet tasks is preserved by the morphisms _ $\parallel$ _. In this sense, categorical representations are data independent descriptions. Thus the same description may apply to whatever can be seen as BioChipNet tasks in a category.

A category of BioChipNet tasks, which is a fundamental and abstract way to describe tasks of a BioChipNet and their relationships, is composed of

• A set of BioChipNet tasks (also called *objects*) together with

• A set of morphisms (sometimes called *arrows*) _ $\parallel$ _ between BioChipNet tasks.

Morphisms _ $\parallel$ _ are to be composable, that is, if $a \parallel b : a \longrightarrow b$ and $b \parallel c : b \longrightarrow c$, then there is a parallel composition $a \parallel b \parallel c$ such that the laws on associativity in (5.21) and on identity in (5.22) are satisfied.

Category of BioChipNet tasks is thus a directed graph with the parallel composition and identity structure. This leads to the following formal definition of the category, named **uTasks**, of BioChipNet tasks. We name this category **uTasks** to refer to "ubiquitous tasks."

Category **uTasks** is a graph (Tasks,ParRel,s,t) consisting of

- Tasks is the set of tasks (considered as nodes).

- ParRel is the set of parallel compositions (considered as edges).

- s,t: ParRel $\longrightarrow$ Tasks are two maps called *source* (or *domain*) and *target* (or *codomain*), respectively.

such that the following axioms (also called *coherence statements*) hold:

- (Associativity) If $a \parallel b$ and $b \parallel c$ then $(a \parallel b) \parallel c = a \parallel (b \parallel c)$. For notational convenience, this can be written as

$$\frac{a \parallel b \quad \text{and} \quad b \parallel c}{(a \parallel b) \parallel c = a \parallel (b \parallel c)} \tag{5.24}$$

- (Identity) For every task a, there exists a morphism $a \parallel skip$ called the identity morphism for a, such that for every morphism $a \parallel b$, we have

$$a \parallel skip \parallel b = a \parallel b = a \parallel b \parallel skip \tag{5.25}$$

We write $a \parallel b : a \longrightarrow b$ when $a \parallel b$ is in ParRel and $s(a \parallel b) = a$ and $t(a \parallel b) = b$.

Property 1 *The category* **uTasks** *is just a core-to-core network on BioChipNet.*

Property 2 *The category* **uTasks** *is a complete graph.*

5.5.3   Categorical characteristics of BioChipNet tasks

Categorical characteristics are often expressed in terms of commutative diagrams and justifications take the form of *diagram chasing*. Informally, a diagram is a picture of some tasks and morphisms in the category **uTasks**. Formally, a diagram is a graph whose nodes are labeled with tasks of **uTasks** and whose edges are labeled with morphisms of **uTasks** in such a way that source and target tasks of an edge are labeled with source and target tasks of the labeling morphism.

Example:
Laws on symmetry in (5.20), on associativity in (5.21) and on identity in (5.22) are,

respectively represented by the following commutative diagrams:

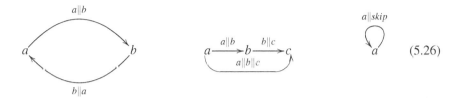

$$(5.26)$$

Example:
The representations in (5.27) include the commutative diagrams of Eq. (5.24).

$$a\|skip \circlearrowright a \xrightarrow{a\|b} b \quad = \quad a \xrightarrow{a\|b} b \quad = \quad a \xrightarrow{a\|b} b \circlearrowleft b\|skip \qquad (5.27)$$

and the equation $(a \| b \| c) \| d = a \| (b \| c \| d)$ is diagrammatically represented as in (5.28)

$$a \xrightarrow{a\|b} b \xrightarrow{b\|c} c \xrightarrow{c\|d} d \underset{a\|b\|c}{} = \quad a \xrightarrow{a\|b} b \xrightarrow{b\|c} c \xrightarrow{c\|d} d \underset{b\|c\|d}{} \qquad (5.28)$$

A path in a diagram is a non-empty sequence of edges and their labeling morphisms such that the target task of each edge is the source task of the next edge in the sequence.

Example:
The central diagram in (5.26) contains the path $a \xrightarrow{a\|b} b \xrightarrow{b\|c} c$, the right diagram in (5.28) contains the paths $a \xrightarrow{a\|b} b \xrightarrow{b\|c} c \xrightarrow{c\|d} d$ and $a \xrightarrow{a\|b} b \xrightarrow{b\|c\|d} d$.

Each path determines a morphism by composing the morphisms along it. A diagram is said to *commute* if, for every pair of tasks x, y, every path from x to y determines the same morphism through composition.

Example:
The following diagram commutes then this amounts to the equation in (5.25) to hold.

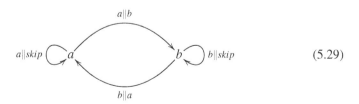

$$(5.29)$$

In a similar way, the following diagram

$$
\begin{array}{cccc}
& a\|b & b\|c & c\|d \\
a & \longrightarrow & b \longrightarrow & c \longrightarrow d
\end{array}
\tag{5.30}
$$

commutes then, in other words, the equation $(a \parallel b \parallel c) \parallel d = a \parallel (b \parallel c \parallel d)$ must hold.

5.5.4 Core-to-core networks

The category **uTasks**, which consists of the set Tasks of tasks together with morphisms _ $\parallel$ _ in the set ParRel, generates core-to-core structure on BioChip-Net. The core-to-core structure is dynamic in nature because tasks can be dynamically added to or dropped from BioChipNet. For such every change, *self-configuration* [310, 312, 313] for the core-to-core structure on BioChipNet occurs.

5.5.4.1 Self-configuration of core-to-core networks

Let $OP = \{add, drop\}$ be the set of operations making a core-to-core structure on BioChipNet change, in which *add* and *drop* are defined as follows:

add is a binary operation

$$
add : \mathsf{ParRel} \times \mathsf{Tasks} \longrightarrow \mathsf{ParRel}
\tag{5.31}
$$

(sometimes specified as $\mathsf{ParRel} \xrightarrow{\;add(\mathsf{Tasks})\;} \mathsf{ParRel}$ or $add(\mathsf{Tasks}) : \mathsf{ParRel} \longrightarrow \mathsf{ParRel}$)

obeying the following axioms: For all $i \in \mathbb{N}_0$,

$$
add(\|_i\, a_i, b) = \begin{cases} (\|_{1 \leqslant i \leqslant n}\, a_i) \parallel b & \text{for } i \geqslant 1 \\ (\|_0) \parallel b = skip \parallel b = b & \text{when } i = 0 \end{cases}
\tag{5.32}
$$

or, also written as

$$
\begin{cases} \|_{1 \leqslant i \leqslant n}\, a_i \xrightarrow{\;add(b)\;} (\|_{1 \leqslant i \leqslant n}\, a_i) \parallel b & \text{for } i \geqslant 1 \\ \|_0 \xrightarrow{\;add(b)\;} (\|_0) \parallel b = skip \parallel b = b & \text{when } i = 0 \end{cases}
$$

or

$$
\begin{cases} add(b) : \|_{1 \leqslant i \leqslant n}\, a_i \longrightarrow (\|_{1 \leqslant i \leqslant n}\, a_i) \parallel b & \text{for } i \geqslant 1 \\ add(b) : \|_0 \longrightarrow (\|_0) \parallel b = skip \parallel b = b & \text{when } i = 0 \end{cases}
$$

Example:

$$add(\|_0, a) \quad = a$$
$$add(a, b) \quad = a \parallel b$$
$$add(a \parallel b, c) = a \parallel b \parallel c$$

drop is a binary operation

$$drop : \mathsf{ParRel} \times \mathsf{Tasks} \longrightarrow \mathsf{ParRel} \qquad (5.33)$$

(sometimes specified as $\mathsf{ParRel} \xrightarrow{drop(\mathsf{Tasks})} \mathsf{ParRel}$ or $drop(\mathsf{Tasks}) : \mathsf{ParRel} \longrightarrow$ ParRel)

obeying the following axioms: For all $i \in \mathbb{N}_0$,

$$drop(\|_i \, a_i, b) = \begin{cases} \|_{1 \leqslant i \leqslant (n-1)} \, a_i & \text{when there exists } a_i = b \\ \|_{1 \leqslant i \leqslant n} \, a_i & \text{for all } a_i \neq b \end{cases} \qquad (5.34)$$

or, also written as

$$\begin{cases} \|_{1 \leqslant i \leqslant n} \, a_i \xrightarrow{drop(b)} \|_{1 \leqslant i \leqslant (n-1)} \, a_i & \text{when there exists } a_i = b \\ \|_{1 \leqslant i \leqslant n} \, a_i \xrightarrow{drop(b)} \|_{1 \leqslant i \leqslant n} \, a_i & \text{for all } a_i \neq b \end{cases}$$

or

$$\begin{cases} drop(b) : \|_{1 \leqslant i \leqslant n} \, a_i \longrightarrow \|_{1 \leqslant i \leqslant (n-1)} \, a_i & \text{when there exists } a_i = b \\ drop(b) : \|_{1 \leqslant i \leqslant n} \, a_i \longrightarrow \|_{1 \leqslant i \leqslant n} \, a_i & \text{for all } a_i \neq b \end{cases}$$

It follows that $drop(\|_0, b) = \|_0 = skip$.

Example:

$$drop(a, a) \qquad = \|_0$$
$$drop(a \parallel b \parallel c, b) = a \parallel c$$
$$drop(a \parallel b \parallel c, d) = a \parallel b \parallel c$$

A process of self-configurations is completely defined when operations *add* and *drop* are executed on a BioChipNet as illustrated in the following diagram:

$$(5.35)$$

In the context of **uTasks**, self-configurations are known as *homomorphisms* from a **uTasks** to another **uTasks** to preserve the core-to-core structure. In other words, self-configuration is a map from a ParRel to another ParRel of the same type that preserves all the parallel structures.

5.5.4.2 Category of core-to-core networks

A category whose objects are categories **uTasks** and whose morphisms are self-configurations is called the category **Cat(uTasks)** of core-to-core networks. The category **Cat(uTasks)** is constructed as follows:

  • *Objects as categories* **uTasks**: Let *ObjCat* be the set of categories **uTasks**. That is,

$$ObjCat = \{\textbf{uTasks} \mid \textbf{uTasks} \text{ is a category of BioChipNet tasks}\} \qquad (5.36)$$

  • *Morphisms as self-configurations*: Associated with each pair of categories **uTasks** and **uTasks**′ in *ObjCat*, self-configuration $h : \textbf{uTasks} \longrightarrow \textbf{uTasks}'$ to map every core-to-core structure to another is a self-configuration from **uTasks** to **uTasks**′ such that for all parallel compositions $a \parallel b$ in ParRel it holds that $h(a \parallel b) = h(a) \parallel h(b)$.

For each pair of self-configurations $h : \textbf{uTasks} \longrightarrow \textbf{uTasks}'$ and $k : \textbf{uTasks}' \longrightarrow \textbf{uTasks}''$, there is an associated self-configuration $h \cdot k : \textbf{uTasks} \longrightarrow \textbf{uTasks}''$, the composition of h with k (and read as "h before k"), such that for all parallel compositions $a \parallel b$ in ParRel it holds that $k(h(a \parallel b)) = k(h(a) \parallel h(b)) = k(h(a)) \parallel k(h(b))$.

Associated with each **uTasks** in *ObjCat*, self-configuration $id_{\textbf{uTasks}} : \textbf{uTasks} \longrightarrow \textbf{uTasks}$ to map every core-to-core structure to itself is an identity self-configuration from **uTasks** to **uTasks** such that for all parallel compositions $a \parallel b$ in ParRel it holds that $id_{\textbf{uTasks}}(a \parallel b) = id_{\textbf{uTasks}}(a) \parallel id_{\textbf{uTasks}}(b) = a \parallel b$.

As a result, for every core-to-core structure and the self-configurations $h : \textbf{uTasks} \longrightarrow \textbf{uTasks}'$, $k : \textbf{uTasks}' \longrightarrow \textbf{uTasks}''$ and $g : \textbf{uTasks}'' \longrightarrow \textbf{uTasks}'''$, the following equations must hold

Associativity:	$(h \cdot k) \cdot g = h \cdot (k \cdot g)$
Identity:	$id_{\textbf{uTasks}} \cdot h = h = h \cdot id_{\textbf{uTasks}'}$

These two equations amount to two following commutative diagrams, respectively.

$$(5.37)$$

and

$$(5.38)$$

These are all the basic ingredients we need for the category **Cat(uTasks)** of core-to-core networks defined.

Property 3 *Every self-configuration* h : **uTasks** $\longrightarrow$ **uTasks**$'$ *and* g : **uTasks**$''$ $\longrightarrow$ **uTasks**$'''$ *in* **Cat(uTasks)**, *there exist unique self-configurations* x : **uTasks**$''$ $\longrightarrow$ **uTasks** *and* y : **uTasks**$'$ $\longrightarrow$ **uTasks**$'''$ *in* **Cat(uTasks)** *such that the following equation holds*

$$x \cdot h \cdot y : \textbf{uTasks}'' \longrightarrow \textbf{uTasks}''' = g : \textbf{uTasks}'' \longrightarrow \textbf{uTasks}''' \quad (5.39)$$

It follows that the self-configurations $id_{\textbf{uTasks}}$: **uTasks** $\longrightarrow$ **uTasks** and $id_{\textbf{uTasks}'}$: **uTasks**$'$ $\longrightarrow$ **uTasks**$'$ are identity self-configurations, then the equation

$$id_{\textbf{uTasks}} \cdot h \cdot id_{\textbf{uTasks}'} : \textbf{uTasks} \longrightarrow \textbf{uTasks}' = h : \textbf{uTasks} \longrightarrow \textbf{uTasks}' \quad (5.40)$$

holds.

Property 4 *For all self-configurations* x *and* y *in* **Cat(uTasks)**, *if* $x \cdot h \cdot y = x \cdot h' \cdot y$, *then* $h = h'$

These properties of self-configurations in **Cat(uTasks)** direct towards a categorical structure, called *extensional monoidal category* as presented in the next section.

5.5.4.3 Extensional monoidal category of core-to-core networks

Further to the category **Cat(uTasks)** of core-to-core networks, we investigate the extensional monoidal structure of the category **Cat(uTasks)** of core-to-core networks. The operation "$\cdot$" defines an extensional monoidal structure on the category **Cat(uTasks)**. In fact, **Cat(uTasks)** equipped with the following multifunctor defines an extensional monoidal category. (Note that we name a monoidal category "extensional monoidal category" when it is equipped with a *multifunctor*, in general, for a distinction from a normal monoidal category just with a *bifunctor*)

$$\cdot : \textbf{Cat(uTasks)} \times \textbf{Cat(uTasks)} \times \textbf{Cat(uTasks)} \longrightarrow \textbf{Cat(uTasks)} \quad (5.41)$$

which, called composition operation, is associative up to a natural isomorphism, and an identity self-configuration id which is both a left and right identity for the multifunctor "$\cdot$," again, up to natural isomorphism. The associated natural isomorphisms are subject to some coherence conditions which ensure that all the relevant diagrams commute. We consider the facts in detail as below.

Property 5 *The composition operation* "$\cdot$" *is associative up to three natural isomorphism* α, β, γ, *called associative ones, with components:*

$$\alpha(h,k,g,d,e) : (h \cdot k \cdot g) \cdot d \cdot e \longleftrightarrow h \cdot (k \cdot g \cdot d) \cdot e$$
$$\beta(h,k,g,d,e) : h \cdot (k \cdot g \cdot d) \cdot e \longleftrightarrow h \cdot k \cdot (g \cdot d \cdot e)$$
$$\gamma(h,k,g,d,e) : (h \cdot k \cdot g) \cdot d \cdot e \longleftrightarrow h \cdot k \cdot (g \cdot d \cdot e)$$

Sometimes, the natural isomorphisms α, β, γ *are also represented as*

$$\alpha(h,k,g,d,e) : (h \cdot k \cdot g) \cdot d \cdot e \cong h \cdot (k \cdot g \cdot d) \cdot e$$
$$\beta(h,k,g,d,e) : h \cdot (k \cdot g \cdot d) \cdot e \cong h \cdot k \cdot (g \cdot d \cdot e)$$
$$\gamma(h,k,g,d,e) : (h \cdot k \cdot g) \cdot d \cdot e \cong h \cdot k \cdot (g \cdot d \cdot e)$$

The coherence conditions for three natural isomorphisms α, β and γ are thought of as diagram (5.45) commuting for all self-configurations h, k, g, d, e, f and p in **Cat(uTasks)**.

Property 6 *Every self-configuration* h : **uTasks** $\longrightarrow$ **uTasks**$'$ *has* $id_{\textbf{uTasks}}$ *and* $id_{\textbf{uTasks}'}$ *as left and right identity self-configurations, respectively. There is a natural isomorphism* λ, *called identity one, with components:*

$$\lambda(h) : id_{\textbf{uTasks}} \cdot h \cdot id_{\textbf{uTasks}'} \longleftrightarrow h \quad or \quad \lambda(h) : id_{\textbf{uTasks}} \cdot h \cdot id_{\textbf{uTasks}'} \cong h \quad (5.42)$$

The coherence condition for the identity natural isomorphism λ is considered as diagram (5.43) commuting for all self-configurations h, k and g in **Cat(uTasks)**.

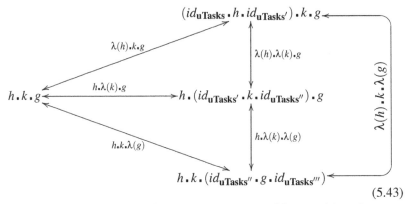

$$(5.43)$$

     The coherence condition states that two or more natural isomorphisms between two given multifunctors are equal based on the existence of which is given or follows from general characteristics. Such situations are ubiquitous in tasks parallel of BioChipNets. Coherence conditions are formulated and studied in categorical structures of BioChipNet tasks parallel known as a categorical approach of the BioChipNet tasks parallel processing.

5.5.4.4   Pushout of self-configuring core-to-core networks

We firstly consider an illustration that if a core-to-core structure is $a \parallel b$ then $a \parallel$ $b \xrightarrow{add(c)} a \parallel b \parallel c$ and $a \parallel b \xrightarrow{drop(b)} a$ are two typical self-configurations. This can be specified by a diagram (see (5.44)) consisting of two self-configurations $add(c) : a \parallel$

$b \longrightarrow a \parallel b \parallel c$ and $drop(b) : a \parallel b \longrightarrow a$ with a common domain.

$$a \parallel b \xrightarrow{\quad add(c) \quad} a \parallel b \parallel c \qquad (5.44)$$

$$\left\downarrow drop(b)\right.$$

$$a$$

Then applying $drop(b)$ on $a \parallel b \parallel c$ and $add(c)$ on a we reach the same structure $a \parallel c$ as described in diagram (5.46).

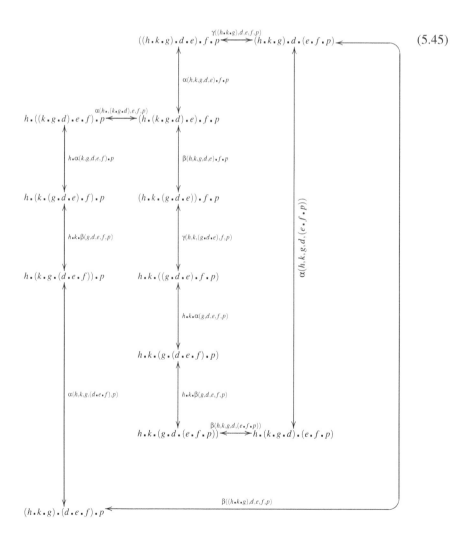

$$(5.45)$$

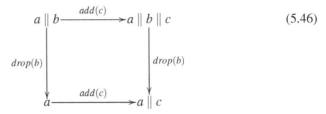

$$(5.46)$$

A *pushout* of a pair of self-configurations $add(c)$: $a \parallel b \longrightarrow a \parallel b \parallel c$ and $drop(b)$: $a \parallel b \longrightarrow a$ is a core-to-core structure $a \parallel c$ together with a pair of self-configurations $drop(b)$: $a \parallel b \parallel c \longrightarrow a \parallel c$ and $add(c)$: $a \longrightarrow a \parallel c$ such that diagram (5.46) commutes.

Moreover, the pushout $\langle a \parallel c, drop(b) : a \parallel b \parallel c \longrightarrow a \parallel c, add(c) : a \longrightarrow a \parallel c \rangle$ must be *universal* with respect to this diagram. That is, for any other pushout such as $\langle b \parallel c, drop(a) : a \parallel b \parallel c \longrightarrow b \parallel c, add(b).add(c).drop(a) : a \longrightarrow b \parallel c \rangle$ there exists a unique self-configuration $drop(a).add(b) : a \parallel c \longrightarrow b \parallel c$ making the following diagram commutes:

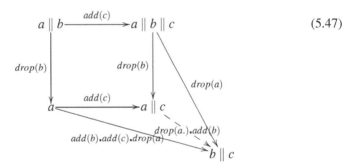

$$(5.47)$$

As with all universal constructions, this pushout is unique up to a bijective self-configuration, i.e. isomorphism.

5.6   Processing BioChipNet data

In this section, we present type of task, type based-agent, category of data types. Then extensional monoidal structure and symmetry monoidal structure of the category of data types are constructed in order to consider the significant properties of data parallel on BioChipNets.

5.6.1   BioChipNet agents

We want to use BioChipNets to formalize data parallel when their tasks interact with each other through input and output. To this end, we put some additional structure on the tasks. Let X and Y be any sets of inputs and outputs, respectively; we define $X \longrightarrow Y$ as *type* of a task on BioChipNet. A task a of the type $X \longrightarrow Y$ is called an *agent* denoted by $a : X \longrightarrow Y$ or $X \xrightarrow{a} Y$. This is read as the task a has inputs from the set X and outputs to the set Y and pictorially drawn as $X \longrightarrow \boxed{\quad a \quad} \longrightarrow Y$.

As suggested by the arrow notation for types, agents can be composed. In fact, if $a : X \longrightarrow Y$ and $b : Y \longrightarrow Z$, then the sequential composition $a \bullet b : X \longrightarrow Z$ is completely defined as follows:

Let A be the set of agents and IO the set of sets of inputs and outputs, then the sequential composition "$\bullet$" is defined by the following functions:

$$
\bullet \overset{def}{=} \left(\begin{array}{ll}
\_ \bullet \_ : A \times A \longrightarrow A & \text{Sequential composition function} \\
dom, cod : A \longrightarrow IO & \text{Domain and codomain functions} \\
id : IO \longrightarrow A & \text{Identity function} \\
\\
\text{such that} \\
\\
dom \; (\_ \bullet \_) = \{f, g : A \mid cod \; f = dom \; g\} \\
\forall f, g, h : A \mid cod \; f = dom \; g \wedge cod \; g = dom \; h \bullet (f \bullet g) \bullet h = f \bullet (g \bullet h) \\
\forall f, g : A \mid cod \; f = dom \; g \bullet dom \; (f \bullet g) = dom \; f \wedge cod \; (f \bullet g) = cod \; g \\
\forall X : IO \; \bullet dom \; (id \; X) = cod \; (id \; X) = X \\
\forall f : A \; \bullet id \; (dom \; f) \bullet f = f \bullet id \; (cod \; f) = f
\end{array} \right)
$$

Note that the identity function $id \; X$ defines an identity agent, denoted as id_X, in A. The identity agent $id_X : X \longrightarrow X$ is pictorially drawn as $X \longrightarrow \boxed{\quad id_X \quad} \longrightarrow X$.

Property 7 *Sequential composition "$\bullet$" is associative.*

Formally, this means that if $a : X \longrightarrow Y$, $b : Y \longrightarrow Z$ and $c : Z \longrightarrow T$ are agents with their types then, by repeated composition, we can construct an agent with type $X \longrightarrow T$ in two ways:

$$(a \bullet b) \bullet c : X \longrightarrow T \qquad \text{and}$$
$$a \bullet (b \bullet c) : X \longrightarrow T$$

So the following equation, also known as a coherence statement, must now hold

$$(a \bullet b) \bullet c : X \longrightarrow T = a \bullet (b \bullet c) : X \longrightarrow T$$

We can see that agents, as described so far to be morphisms, do not form a category of data types yet under sequential composition "$\bullet$." The reason is that there are no identity morphisms. However, we will obtain such a category in Section 5.6.2.

5.6.2 Category of BioChipNet data types

By the defined structure of agents, we can construct **uData** to be a category of data types. In fact, **uData** is constructed as follows:

• *Objects as the sets of data*: Let *IO* be the set, whose elements are the sets of inputs to or outputs from tasks. That is,

$$IO = \{X \mid X \text{ is a set of inputs to or outputs from tasks}\} \qquad (5.48)$$

• *Morphisms as agents*: Let *A* be the set of agents. Then associated with each objects *X* in *IO*, we define morphism from *X* to *X*, called identity morphism on *X*, as an agent $X \xrightarrow{id_X} X$ in *A* or

This means as

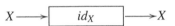

and to each pair of agents $X \xrightarrow{a} Y$ and $Y \xrightarrow{b} Z$, there is an associated agent $X \xrightarrow{a \cdot b} Z$, the composition of *a* with *b*. This means that if we have two agents $X \longrightarrow \boxed{a} \longrightarrow Y$ and $Y \longrightarrow \boxed{b} \longrightarrow Z$ then there exists the associated agent

$$X \longrightarrow \boxed{a \cdot b} \longrightarrow Z = X \longrightarrow \boxed{a} \xrightarrow{Y} \boxed{b} \longrightarrow Z$$

For every object X, Y, Z and T in *IO* and the agents $X \xrightarrow{a} Y$, $Y \xrightarrow{b} Z$ and $Z \xrightarrow{c} T$ in the set *A* of agents, then the following equations, also known as the coherence statements, hold:

Associativity:	$(a \cdot b) \cdot c : X \longrightarrow T = a \cdot (b \cdot c) : X \longrightarrow T$
Identity:	$id_X \cdot a : X \longrightarrow Y = a : X \longrightarrow Y = a \cdot id_Y : X \longrightarrow Y$

In other words, these coherence statements can be diagrammatically drawn such as

$$(X \longrightarrow \boxed{a \cdot b} \longrightarrow Z) \longrightarrow \boxed{c} \longrightarrow T =$$
$$X \longrightarrow \boxed{a} \longrightarrow Y(\longrightarrow \boxed{b \cdot c} \longrightarrow T)$$

and

$$X \longrightarrow \boxed{id_X} \longrightarrow X \longrightarrow \boxed{a} \longrightarrow Y =$$
$$X \longrightarrow \boxed{a} \longrightarrow Y =$$
$$X \longrightarrow \boxed{a} \longrightarrow Y \longrightarrow \boxed{id_Y} \longrightarrow Y$$

These are all the basic ingredients we need to form the category, named **uData**, of data types. Agents are closed under the sequential composition ".". Moreover, the agents such as $id_X : X \longrightarrow X$ act as identity morphisms for this composition.

Thus, category **uData** has sets (denoted by X, Y etc, for example) as its objects, and agents (e.g., $a : X \longrightarrow Y$) as its morphisms.

Property 8 *Every agent $a : X \longrightarrow Y$ and $c : Z \longrightarrow T$ in* **uData**, *there exist unique agents $x : Z \longrightarrow X$ and $y : Y \longrightarrow T$ in* **uData** *such that the following equation holds*

$$x \cdot a \cdot y : Z \longrightarrow T = c : Z \longrightarrow T \tag{5.49}$$

It follows that the agents $id_X : X \longrightarrow X$ and $id_Y : Y \longrightarrow Y$ are identity agents, then the equation

$$id_X \cdot a \cdot id_Y : X \longrightarrow Y = a : X \longrightarrow Y \tag{5.50}$$

holds.

Property 9 *For all agents x and y in* **uData**, *if $x \cdot a \cdot y = x \cdot b \cdot y$, then $a = b$*

These properties of agents in **uData** lead to a categorical structure, called *extensional monoidal category* as below.

5.6.3　Extensional monoidal category of BioChipNet data types

The composition operation "." defines an extensional monoidal category on the category **uData**. In fact, the extensional monoidal category is a category **uData** equipped with the following additional structure:

- A multifunctor $. :$ **uData** $\times$ **uData** $\times$ **uData** $\longrightarrow$ **uData** called the composition operation

- For every agent $a : X \longrightarrow Y$, id_X and id_Y called left and right identity agent, respectively

- The natural isomorphisms α, β, γ and λ subject to some coherence conditions expressing the fact that

  1. *The composition operation* "." *is associative*: there are three natural isomorphisms α, β, γ, called associative ones, with components

$$\alpha(a,b,c,d,e) : (a \cdot b \cdot c) \cdot d \cdot e \longleftrightarrow a \cdot (b \cdot c \cdot d) \cdot e$$
$$\beta(a,b,c,d,e) : a \cdot (b \cdot c \cdot d) \cdot e \longleftrightarrow a \cdot b \cdot (c \cdot d \cdot e)$$
$$\gamma(a,b,c,d,e) : (a \cdot b \cdot c) \cdot d \cdot e \longleftrightarrow a \cdot b \cdot (c \cdot d \cdot e)$$

Sometimes, the natural isomorphisms α, β, γ are also represented as

$$\alpha(a,b,c,d,e) : (a \cdot b \cdot c) \cdot d \cdot e \cong a \cdot (b \cdot c \cdot d) \cdot e$$
$$\beta(a,b,c,d,e) : a \cdot (b \cdot c \cdot d) \cdot e \cong a \cdot b \cdot (c \cdot d \cdot e)$$
$$\gamma(a,b,c,d,e) : (a \cdot b \cdot c) \cdot d \cdot e \cong a \cdot b \cdot (c \cdot d \cdot e)$$

2. *The composition operation "$\cdot$" has id_X and id_Y as left and right identities of every agent $a : X \longrightarrow Y$: there is a natural isomorphism λ, called identity one, with components $\lambda(a) : id_X \cdot a \cdot id_Y \longleftrightarrow a$ or $\lambda(a) : id_X \cdot a \cdot id_Y \cong a$*

The coherence conditions for three natural isomorphisms α, β and γ are thought of as diagram (5.51) commuting for all agents a, b, c, d, e, f and g in **uData**.

The coherence condition for the identity natural isomorphism λ is seen as diagram (5.52) commuting for all agents $a : X \longrightarrow Y, b : Y \longrightarrow Z$ and $c : Z \longrightarrow T$ in **uData**.

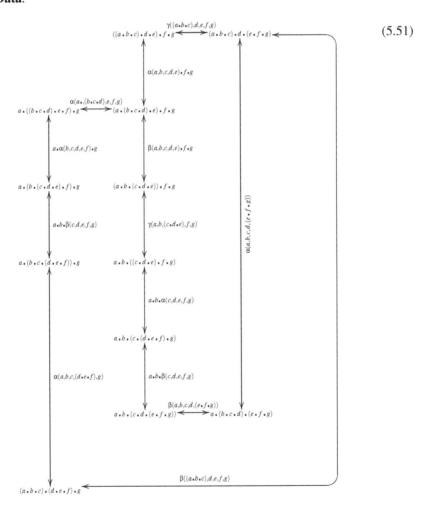

(5.51)

The coherence condition expresses the statement that two or more natural isomorphisms between two given multifunctors are equal based on the existence of which is given or follows from general characteristics. Such situations are ubiquitous in data parallel of BioChipNets. Coherence conditions are formulated and studied in categorical structures of BioChipNet data parallel known as a categorical approach of the BioChipNet data parallel processing.

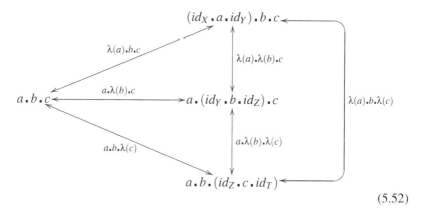

$$(5.52)$$

5.6.4 Parallel composition of BioChipNet agents

What other operations, besides sequential composition ".," do we have on agents? To begin with, there is an obvious notion of parallel composition. Let $X + X'$ denote the disjoint union of sets X and X'. Then given two agents $a : X \longrightarrow Y$ and $b : X' \longrightarrow Y'$, we can define an agent $a \parallel_a b : X + X' \longrightarrow Y + Y'$ as follows:

$$\parallel_a \stackrel{def}{=} \left(\begin{array}{l} \_ + \_ : IO \times IO \longrightarrow IO \\ \_ \parallel_a \_ : A \times A \longrightarrow A \\ id_{\parallel_a} : A \\[2mm] \text{such that} \\[2mm] \forall f, g, h : \Lambda \bullet (f \parallel_a g) \parallel_a h - f \parallel_a (g \parallel_a h) \\ \forall f : A \bullet f \parallel_a id_{\parallel_a} = id_{\parallel_a} \parallel_a f = f \\ \forall X, Y : IO \bullet X + Y = dom(id\, X \parallel_a id\, Y) \\ \forall X, Y : IO \bullet id\, X \parallel_a id\, Y = id(X + Y) \\ \forall f, g, p, q : A \mid cod\, f = dom\, g \wedge cod\, p = dom\, q \bullet (f \bullet g) \parallel_a (p \bullet q) = \\ (f \parallel_a p) \bullet (g \parallel_a q) \end{array} \right)$$

A diagram of parallel composition $\parallel_a$ is pictorially represented as in diagram (5.54). The operation $\parallel_a$ defines a symmetric monoidal structure on the category **uData**. In fact, **uData** equipped with the following bifunctor defines a symmetric monoidal category

$$+ : \mathbf{uData} \times \mathbf{uData} \longrightarrow \mathbf{uData} \qquad (5.53)$$

which, called parallel composition operation, is associative up to a natural isomorphism, and an empty set $\emptyset$ which is both a left and right identity for the bifunctor "+," again, up to natural isomorphism. The associated natural isomorphisms are subject to some coherence conditions which ensure that all the relevant diagrams commute. We consider the facts in detail as below.

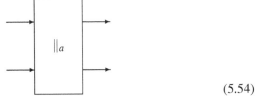

$$(5.54)$$

Property 10 *The parallel composition operation "$+$" is associative up to a natural isomorphism* α_+, *called associativity, with components:*

$$\alpha_+(X,X',X'') : (X+X')+X'' \longleftrightarrow X+(X'+X'') \qquad (5.55)$$

or, sometimes, this associativity is also written as

$$\alpha_+(X,X',X'') : (X+X')+X'' \cong X+(X'+X'')$$

The coherence condition for the associativity α_+ means that for all sets of inputs or outputs X, X', X'' and X''' in *IO*, diagram (5.56) commutes.

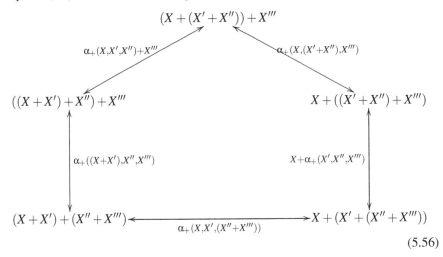

$$(5.56)$$

Property 11 *The parallel composition operation "$+$" has an identity $\emptyset$ up to left identity natural isomorphism* λ_+ *and right identity natural isomorphism* ρ_+ *with components:*

$$\lambda_+(X) : \emptyset+X \longleftrightarrow X \qquad (5.57)$$
$$\rho_+(X) : X+\emptyset \longleftrightarrow X \qquad (5.58)$$

or, sometimes, these left and right identity natural isomorphisms are also written as

$$\lambda_+(X) : \emptyset+X \cong X$$
$$\rho_+(X) : X+\emptyset \cong X$$

Moreover, the coherence conditions for the left and right identity natural isomorphisms λ_+ and ρ_+ mean that for all sets of inputs or outputs X and X' in IO, diagram (5.60) commutes.

Property 12 *Given agents* $b : X \longrightarrow X'$, $c : X' \longrightarrow X''$, $p : Y \longrightarrow Y'$ *and* $q : Y' \longrightarrow Y''$, *then two following expressions are equal:*

$$(b \cdot c) \|_a (p \cdot q) = (b \|_a p) \cdot (c \|_a q) \tag{5.59}$$

where both left and right expressions define an agent, whose type is $X + Y \longrightarrow X'' + Y''$

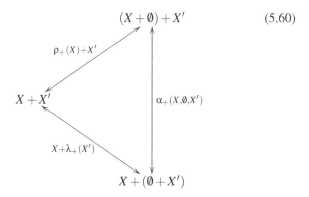

$$(5.60)$$

Property 13 *Given the identity agents* $id_X : X \longrightarrow X$ *and* $id_{X'} : X' \longrightarrow X'$, *then*

$$id_X \|_a id_{X'} = id_{X+X'} \tag{5.61}$$

where $id_{X+X'}$ *is the identity agent* $id_{X+X'} : X + X' \longrightarrow X + X'$

In the category **uData**, we have $X + X' = X' + X$ for the sets of X and X' in IO. In particular, if the agent $id_{X+X'} : X + X' \longrightarrow X + X'$ then it is equivalent to $id_{X+X'} : X + X' \longrightarrow X' + X$ called a symmetric agent between $X + X'$ and $X' + X$. This means that **uData** is what is called a symmetric monoidal category.

Here, as a practice offered to readers, let us try to represent the coherence condition for the symmetric agent $id_{X+X'}$. In other words, we have to draw commutative diagram(s) for $id_{X+X'}$ in **uData**.

Formally, for X and X' in IO, the symmetric agent $id_{X+X'}$ is the natural isomorphism between $X + X'$ and $X' + X$, which is defined as follows:

$$id_{X+X'} \stackrel{def}{=} \left(\begin{array}{l} id_{\_+\_} : IO \times IO \longrightarrow A \\ id : IO \longrightarrow A \\ \\ \text{such that} \\ \\ \forall X, X' : IO \bullet dom(id_{X+X'}) = X + X' \wedge cod(id_{X+X'}) = X' + X \\ \forall X, X' : IO \bullet id_{X+X'} \bullet id_{X'+X} = id\ (X + X') \\ \forall X, X', X'' : IO \bullet id_{(X+X')+X''} = \\ \qquad\qquad (id\ X \parallel_a id_{X'+X''}) \bullet (id_{X+X''} \parallel_a id\ X') \end{array} \right)$$

5.7 Notes and remarks

The aim of this chapter has been both to give an in-depth analysis as well as to present the new material on the applied categorical structures for both tasks parallel and data parallel on BioChipNets for firmly supporting the core-to-core and agent-based networking techniques out of the ANoC techniques such as grid, cluster, ubiquitous [7, 14, 15]. Unlike the traditional parallel processing for chips based on multiple processors, processing and developing parallel computations and communications on such BioChipNets are raising challenges not tackled before, including *nondeterminism, decentralization, context-awareness and goal-and inference-driven adaptability* [118, 129, 175, 185]. Hence, in future, we hope to address further developments which are not exposed yet in this chapter. However, below we briefly discuss the direction of these developments.

For the category **Cat(uTasks)** of core-to-core networks, we develop a category **Cat(uTasks)**$'$ equipped with the following structures:

- Let $Obj(\mathbf{Cat(uTasks)}')$ be the set of core-to-core networks **uTasks**. Every **uTasks** is called an object of **Cat(uTasks)**$'$.

- For each pair $(\mathbf{uTasks}, \mathbf{uTasks}')$ of objects of **Cat(uTasks)**$'$, let *Self-* $*(\mathbf{uTasks}, \mathbf{uTasks}')$ be an object of **Cat(uTasks)**, called the *(Self-*)-object* of **uTasks** and **uTasks**$'$.

- For each object **uTasks** of **Cat(uTasks)**$'$, let $id(\mathbf{uTasks})$ be a morphism in **Cat(uTasks)** from $id_{\mathbf{uTasks}}$ to *Self-*$*(\mathbf{uTasks}, \mathbf{uTasks})$, called the *identity morphism* of **uTasks**. That is

$$id_{\mathbf{uTasks}} \xrightarrow{\ id(\mathbf{uTasks})\ } \textit{Self-}*(\mathbf{uTasks}, \mathbf{uTasks})$$

- For each triple (**uTasks**, **uTasks′**, **uTasks″**) of objects of **Cat(uTasks)′**, let

$$Self\text{-}*(\mathbf{uTasks}, \mathbf{uTasks'}) \times Self\text{-}*(\mathbf{uTasks'}, \mathbf{uTasks''})$$
$$\downarrow \text{``.''}$$
$$Self\text{-}*(\mathbf{uTasks}, \mathbf{uTasks''})$$

 be a morphism in **Cat(uTasks)** called the composition morphism of **uTasks**, **uTasks′** and **uTasks″**.

If all *Self-**(**uTasks**, **uTasks′**) satisfy three axioms including associativity, left identity and right identity then **Cat(uTasks)′** is a category *enriched* over **Cat(uTasks)**.

For the category **uData** of data types, we also develop a category **uData′** equipped with the following structures:

- Let *IO* be the set of inputs or outputs. Every element X in *IO* is called an object of **uData′**.

- For each pair (X,Y) of objects of **uData′**, let $agent(X,Y)$ be an object of **uData**, called the *agent-object* of X and Y.

- For each object X of **uData′**, let $id(X)$ be a morphism in **uData** from id_X to $agent(X,X)$, called the *identity morphism* of X. That is

$$id_X \xrightarrow{\ id(X)\ } agent(X,X)$$

- For each triple (X,Y,Z) of objects of **uData′**, let

$$agent(X,Y) \times agent(Y,Z)$$
$$\downarrow \text{``.''}$$
$$agent(X,Z)$$

 be a morphism in **uData** called the composition morphism of X, Y and Z.

If all $agent(X,Y)$ satisfy three axioms including associativity, left identity and right identity then **uData′** is a category *enriched* over **uData**.

Despite many significant successes, middleware development for ANoC is still highly challenging. In fact, ANoC paradigm is reflected by several forms of communicating such as grid, ubiquitous, cluster, core-to-core, agent-based networking and so on. The core-to-core and agent-based networking have formally been exhibited in this chapter. For rigorously approaching to the other mentioned forms of ANoC, then algebraic extensions over the notions of tasks parallel and data parallel on BioChip-Nets, as open problems for future work, are needed.

5.8   Conclusions

In this chapter, based on categorical structures, we have rigorously approached to the aspects of tasks parallel and data parallel on bio-inspired on-chip networks (BioChip-Nets) from which their useful properties emerge.

For tasks parallel, we have started with investigating the parallel composition of tasks $\|_{i \in \mathbb{N}_0} a_i$ on BioChipNets to construct every core-to-core network as a category **uTasks** of BioChipNet tasks together with the notion of diagram chasing, then the extensional monoidal category **Cat(uTasks)** of such core-to-core networks together with categorical aspects of self-configuration has been developed in detail.

For data parallel, our developments have stemmed from formulating type of tasks as $X \longrightarrow Y$, type based-agent as $a : X \longrightarrow Y$ and category **uData** of data types. Then extensional monoidal structure and symmetry monoidal structure of the category of data types have been constructed in order to consider the significant properties of data parallel on BioChipNets.

A direction of our further developments has also been investigated including *enriched structures* **Cat(uTasks)**$'$ over the extensional monoidal category **Cat(uTasks)** of core-to-core networks and **uData**$'$ over the extensional monoidal category **uData** of data types. Moreover, algebraic extensions over the notions of tasks parallel and data parallel on BioChipNets are needed for rigorously approaching to the other forms of ANoC involving grid, ubiquitous, cluster networking.

6

HAMSoC: A Monitoring-Centric Design Approach for Adaptive Parallel Computing

L. Guang

Department of Information Technology, University of Turku, Finland

J. Plosila

Department of Information Technology, University of Turku, Finland

J. Isoaho

Department of Information Technology, University of Turku, Finland

H. Tenhunen

Department of Information Technology, University of Turku, Finland

CONTENTS

Adaptive complex parallel/distributed embedded system is an appealing dream, but without a suitable design methodology, the dream may be unattainable. We propose a novel design approach, Hierarchical Agent Monitored System-on-Chip (HAMSoC), which exploits design separation and abstraction to provide scalable and efficient design of monitoring functions. Here we focus on presenting the formal specification and modeling of the design platform, which provides a proper HW/SW codesign abstraction required for VLSI systems. A design example is given on hierarchical power monitoring for Network-on-Chip (NoC).

6.1   Introduction

Constant technology scaling has enabled parallel many-core System-on-Chip (SoC) to become a mainstream general-purpose computing platform. Tilera processor [26] has 64 cores fabricated in 90nm technology, and TeraFLOPS [305] has 80 cores in 65nm technology. Embedded system in general will expand both in the integration on-chip and into wider physical spaces with hybrid manners of communication methods. For instance, Cyber-Physical System (CPS) is projected to appear, in the domain of embedded computing, as networked embedded computers monitoring physical processes [174]. Parallel and distributed embedded system is an inevitable design trend in the future.

For on-chip integration, one major challenge for future Ultra-Large-Scale-Integration (ULSI) electronic system comes from the stronger influence of variations, including PVT (process, voltage and thermal) variations as well as physical variations caused by aging [33]. The relative deviation proportions of transistor and wire parameters on manufactured chips become larger as the nominal parameters become smaller. In addition, submicron effects, such as crosstalk, affect circuit and system performance significantly, as a result of shrinking distance between devices and interconnects.

Adaptivity is an essential system capability to deal with the challenges of emerging massively parallel on-chip systems or distributed embedded systems in general. Adaptive (re)configuration not only tackles variations but also exploits variations to improve performance or efficiency. For instance, Razor [75] architecture exploits the design margin between the worst-case voltage supply and the minimal value required for a given working frequency. By dynamically observing the timing errors in circuits, the V_{DD} (supply voltage) is adaptively adjusted at the minimal values without timing errors, so that the system gains significant power efficiency compared to using static power supplies. [252] generalizes such design method as Always-Optimal-Design, and the way to achieve the highest efficiency is to adaptively adjust system modules based on "operating, manufacturing and environmental conditions."

To provide adaptivity on massively parallel systems, on-chip or distributed, designers face the overwhelming pressure of system complexity with the opposite demands of shorter time-to-market. Design separation and abstraction are two important principles to deal with complexity, as we can observe in the persistent research in design methodology [153]. We will elaborate the background of design methodology in Section 6.2.1. As an important tool in providing efficient and dependable design of complex parallel system, formal methods are utilized in various stages of the design process. The formal specification is consistently validated from system-level design to implementation, and formal modeling is used for analyzing and validating functions and performance.

This chapter presents an innovative design approach, Hierarchically Agent Monitored Parallel Computing. It follows monitoring-centric design methodology, which separates and concentrates on the design of monitoring functions. We consider monitoring as a generalized term for all necessary control and adjustment processes in the system. Hierarchical Agent Monitored SoC is the application of the design approach on massively parallel on-chip systems. A formal specification framework is provided, which captures the system-level design with proper abstraction so that the design effort is minimized. FSM (Finite-State Machine)-based modeling is presented for early-stage monitoring-triggered state analysis. As Network-on-Chip becomes a widely adopted parallel on-chip computing platform, and power management is one of the most needed monitoring services, a design example of hierarchical monitoring operations on Network-on-Chip is demonstrated.

The rest of the chapter is organized as follows: Section 6.2 presents the HAMSoC design approach. Section 6.3 elaborates on the formal specification and modeling of the design platform. Section 6.4 presents the case study on NoC platform. Section 6.5 concludes the chapter.

6.2 Hierarchical agent monitoring design approach

Hierarchical agent monitoring design approach adopts a new design concentration, monitoring-centric design. We will first look at the motivation of the novel methodology (Section 6.2.1), and continue to present hierarchical agent monitoring design approach (Section 6.2.2), in particular on SoC platforms (Section 6.2.3).

6.2.1 Monitoring-centric design methodology

Design separation and abstraction are important design principles. Design separation is also called design orthogonalization [153], as we concentrate on the design concerns of particular aspects. It by no means implies that various aspects of the design process are not interleaved, instead it motivates the concentration of specific design aspects to improve efficiency and portability. Current design space exploration mainly consists of two dimensions: computation and communication. The communication-centric design was proposed when the parallel computing became so complicated that traditional ad-hoc communication design was no longer suitable [253]. The design concentration of communication architecture enables more ef-

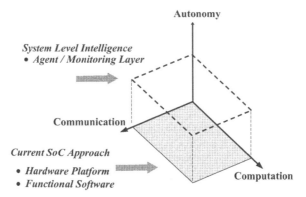

FIGURE 6.1
3-Dimensional design space for parallel embedded systems.

ficient design, analysis and validation of various communication architectures, which can be ported to different application scenarios.

These design principles have been captured in existing design methodologies to deal with system complexity. Model-based design [141, 267] abstracts the components (processing elements and control logics) into models with functional, timing and communication specification. It was the initial design revolution towards efficient design reuse and easier verification at an early design stage. Platform-based design [153, 265] provides a higher-level abstraction. Conceptually we can perceive the platform as an abstraction layer which captures the functional and non-functional features of a particular design level. The definition of platforms varies significantly based on the actual system, or the level designers work on. The most general platforms are software, hardware and API (application programming interface) platforms [153, 265].

Here we propose monitoring-centric design, as a new dimension in addition to current design concentration of computation and communication (Figure 6.1). This dimension focuses on the design, implementation and application of monitoring components and algorithms, to achieve system adaptivity and autonomy. It brings a new design platform for monitoring services, considered from the perspective of platform-based design. Monitoring operations are needed for any distributed system in general, but for embedded systems, monitoring-centric design has its special features since embedded computing is resource limited. We can reason the necessity of monitoring-centric design from several aspects.

First, monitoring operations have become critical in complex embedded systems. Scaling towards technology nodes of 45nm and beyond, we can observe the dramatic increase of transistor and circuit level undependability [272], and components are more likely to suffer from errors [43]. In order to build dependable system, run-time errors and variations need to be dealt with properly. Conventional worst-case design brings non-affordable overhead in terms of silicon area and power, which can be greatly reduced if dynamic monitoring can be realized. For instance, Razor architecture [75] reports significant power saving compared to either static or conventional monitoring method (dynamic voltage scaling without exploiting the process variation).

Moreover, the design and implementation of monitoring operations have specific concerns and tradeoff, which are different from the design of computation and communication. This issue is clearer when we focus on the embedded system design, where the resource limitation raises numerous design concerns for monitoring operations. For example, fine-grained DVFS (dynamic voltage and frequency scaling) is an advanced power supply management methods for power saving [157]. The design and implementation of monitoring circuitry for the involved reconfiguration process pose several challenges. For instance, on-chip converters bring considerable timing delay and area overhead [157] despite the effort of circuit designers. Without fast and low-overhead voltage switching, fine-grained power management with traffic condition monitoring is meaningless. Awareness of this design concern motivates the proposal of multiple on-chip delivery networks for voltage switching [296], which provides much lower switching delay and far better scalability than on-chip DC converters.

In addition, design concentration on monitoring components and operations will enable portable design across platforms, just as the portable communication architectures identified in communication-centric design. For instance, the previously mentioned Razor architecture was originally proposed on processing elements, its monitoring principle can be applied on interconnection as well [323].

6.2.2   Hierarchical agent monitoring

Hierarchical agent monitoring design approach, following the monitoring-centric design methodology, abstracts the monitoring functions into a design concept, "agent." It can be any monitoring entity which is able to perform certain static or adaptive monitoring operations. Conceptually, the operation of an agent can be illustrated in Figure 6.2. In terms of functions, an agent is observing and configuring certain resources assigned to it. It traces the parameter of the resources, while receiving messages from other agents. The agent processes these information using built-in cost functions and determines if and how the reconfiguration should be performed on the resources. It may also send messages to other agents, in order to share information or issue configuration commands. The term "agent" originally comes from artificial intelligence and software engineering, with a wide diversity of meanings and interpretation [100]. One classic definition can be found in [262]: "an agent is anything that can be viewed as perceiving its environment through sensors and acting upon that environment through effectors." Clearly, the concept of agent does not limit its implementation into any particular form.

Hierarchical agent monitoring design approach relies on hierarchical agent monitoring architecture. Hierarchical control or monitoring structure, as an advanced parallel system architecture, is acknowledged in existing works. [24] presents a project which develops biologically motivated computing organism. By learning from the

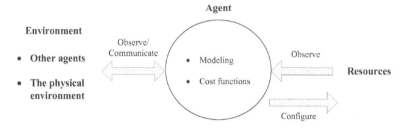

FIGURE 6.2
Conceptual overview of an agent.

multi-level control architecture in biological systems ("brain," "organ," and "cell"), the work argues the integration of multi-level reconfiguration control in computing systems. In addition, decentralized control schemes has been noted for their higher efficiency and lower overhead in many previous works (for instance [95]).

Our design approach provides a systematic method applying hierarchical monitoring structure. The agent hierarchy constitutes a generic design platform suitable for any potential monitoring functions. Agents on each level have clearly defined scopes and priorities (Figure 6.3). The application agent is defined as a dynamic module which sends the application requirements to the platform. Such requirements may be the maximum execution time of the program, or part of the program, or the maximal latency for certain computation or communication flows. Such information is only known to the application designers, and must be informed to the platform. At the platform side, a platform agent is the top level monitor, which tries to configure the platform to meet the requirement sent by the application agent. In other words, the platform (including all types of resources in the platform) is transparent to the application. As the number of components in a system gets larger, the centralized manner of monitoring with only the platform agent becomes inefficient. Thus a number of cluster agents are assigned, by the platform agent, to monitor different regions (clusters) respectively. As the localized monitoring goes further, each cluster will be composed of finer-granular components, or cells, each monitored by a cell agent. A cell is considered atomic in the monitoring operation, as it is configured as a unified entity in the monitoring process.

It should be emphasized that, from the perspective of design methods, the term "agent" is adopted to enable the separation of monitoring functions from computation and communication. We do not study agents from the perspective of artificial intelligence. Instead, we research on how to design and implement adaptive operations on parallel or distributed embedded systems, using the design abstraction. Agents may be running different algorithms, and implemented in different manners. These issues are not specified by the design approach but are determined by the actual system requirements, for instance the on-chip system (Section 6.2.3).

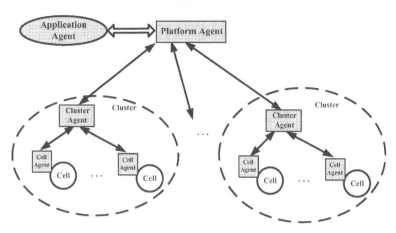

FIGURE 6.3
Hierarchical agent monitored platform.

6.2.3 Hierarchical agent monitored system-on-chip

HAMSoC is the application of the hierarchical agent monitoring approach on massively parallel on-chip systems. When instantiated on SoC platforms, the implementation of agents and their interfaces to other parts of the system become more straightforward due to the modularized parallel SoC architecture (Figure 6.4). A whole agent entity encompasses all components for monitoring operations, including sensors, the monitoring processing logic, communication interface to other agents, and the actuators to apply the reconfiguration. Sensors provide information from the monitored resources, and they can be analog or digital circuits. The monitoring processing logic can be synthesized as software cores (general purpose controllers) or hardware circuits. Actuators are circuits needed to realize the intended reconfiguration, for instance frequency synthesizers or voltage converters.

HAMSoC approach promotes that any monitoring operation should always be put on proper structural level based on its granularity and physical overhead, so that the design can be scalable both in terms of design efficiency as well as physical implementation.

On the platform level, the platform agent performs the global setting, reconfiguration and monitoring. Such monitoring decisions should be based on the information of all available resources, and attempt to achieve the global optimization of system performance. Common global configurations in parallel SoC platforms include network configuration, voltage island partition, dynamic reconfiguration in case of regional errors. For instance, the well-known ElastIC architecture adopts a central diagnostic and adaptivity processing unit (DAP) [292], to supervise a large number of small but simple processing elements. This centralized monitor is designed to perform highly-prioritized operations on any processing elements, including taking them offline for testing, and active healing in case of device errors and performance

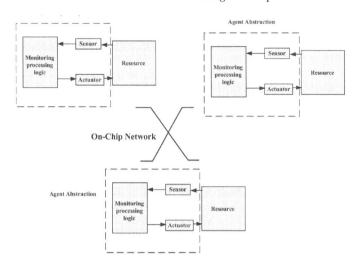

FIGURE 6.4
Typical agent monitoring interface in parallel SoC platforms.

degradation. Global monitoring operations may require a large variety of processing with complex algorithms. In addition, since the run-time information of all resources needs to be used in such processing, the required memory storage is quite high. As a result, the implementation of the platform agent usually contains software-based monitoring processing logic accompanied by hardware circuits as sensors and actuators.

On the cluster level, cluster agent applies the configuration decided by the platform agent within the cluster where it functions as the general monitor. On the emerging massively parallel platforms, different regions may be running various types of processing in an application. For instance, on the 167-core platform introduced in [296], 9 processors are used for JPEG encoder, and at most 20 processors are used for a H.264 decoder. Such integration will become common in future SoC architectures. Thermal management is another example of regional monitoring operation, as the workload imbalance in different areas will lead to thermal distribution with hotspot [277]. The implementation of cluster agent should be based on the actual algorithm complexity. Complicated cluster-level operations shall be handled by software-based controllers, while hardware-based monitors may fulfill relatively simple functionality with lower overhead. For instance, the power management in a 80-tile network-on-chip as in [305] is mostly hardware-based.

On the cell level, cell agents trace local circuit status, such as current, workload and any faults or failures. These operations are usually specified by simple algorithms and require low latency. For instance, communication errors can be detected on the interconnect, and retransmission is a simple fault recovery technique [176]. The detection and recovery of errors on individual links are preferably performed

TABLE 6.1

Hierarchical agent monitoring operations in parallel SoCs

Agent Level	General Function	Feature	Implemen -tation	Example
Platform Agent	global monitoring and configuration	coarse-grained, system influence, rarely issued, long latency	software-dominated	dynamic mapping, dynamic network configuration
Cluster Agent	applying platform agent's configuration, regional monitoring	coarse-grained, only influence local cluster	SW/HW codesign	voltage island, application block mapping
Cell Agent	applying cluster agent's configuration, local monitoring	fine-grain, frequent and high-speed	hardware-dominated	circuit error detection, per-core DVFS, power gating

as fast as possible, to prevent redundant transmission in faulty channels. In addition, such operations are usually supported by dedicated distributed hardware. For instance, the power switches [296] enable speedy reconfiguration to different voltages dynamically. As a result, cell agents are usually hardware monitors responsible for the setting of involved supporting hardware of the monitoring operations.

In addition to monitoring resources, agents are also monitoring lower-level agents. It is important since no matter how adaptive agents are, they are also victims of errors and failures. The failures of agents can be handled in the same way as resource failures by their upper level agents.

Table 6.1 summarizes the functional partition and implementation guidelines of agents in SoC platforms.

In summary, HAMSoC approach provides scalability from the design method's perspective. First, the separation of monitoring operations onto different levels supports focused design efforts on a particular structural level. For massively parallel computing platforms, this design abstraction will greatly speed up the design process. Second, hierarchical monitoring structure can be easily scaled by adding superclusters upon clusters, if the applications running in a system have more complex hierarchy. Third, the physical scalability of the design approach is provided by the operation-specific SW/HW codesign guideline. Any monitoring operation should be synthesized based on the involved complexity. Typically the platform level agent is a software-based controller, whose overhead is incurred once only on the whole system. Meanwhile, the largest number cell agents are commonly hardware circuits with smallest possible overhead. Our previous work [116] gives more elaborated quantitative analysis on the issue for interested readers.

6.3   Formal specification of HAMSoC

Formal methods are widely acknowledged as important and sometimes indispensable in modern embedded system design. Various specification languages, formal models, and validation tools have been proposed and used in every stage of system design and implementation, from system specification and component modeling (software and hardware), to the modeling and validation of low-level circuits. On the system-level design the importance of formal methods has become clear when modern systems get increasingly complex and heterogeneous. System behaviors should be specified and verified based on formal models at a high level of abstraction [86]. In particular, on massively parallel systems with heterogeneity and adaptivity, considering the run-time interactions between parallel units with dynamic reconfiguration, it is difficult to validate the system by exhaustive simulation, if not impossible. For instance, on one of the new computing frontiers, cyber-physical systems (CPS) [173], we are dealing with a potentially very large number of networked, diverse and adaptive distributed devices with various communication channels. As the system complexity and unpredictability become unprecedentedly high, formal modeling will be inevitable in the development process [196].

Formal methods have been integrated from the initial stage of the development of HAMSoC approach, starting from a formal specification of the design platform. We propose a new specification language, in order to enable a highly-abstracted monitoring-centric design. In the specification, the monitoring operations and parameters are directly defined. To support the SW/HW codesign of SoC platforms, the formal specification exposes necessary hardware details for defining monitoring operations. For instance, the specification differentiates three types of parameters, P_{STA} (static parameter), P_{OBS} (observable parameter) and P_{REC} (reconfigurable parameter), in order to capture the alternatives in monitoring and reconfiguration process. To facilitate the early-stage state analysis, FSM (finite-state machine) is adopted in modeling state transition triggered by monitoring operations. The state transition also emphasizes on modeling the actual hardware operation at a high level abstraction. For instance, stable and transitional values are differentiated to account for the reconfiguration delay of certain parameters. The formal specification can be transformed into any suitable generic formal languages for further automated verification.

6.3.1   Specification framework of HAMSoC

The specification framework to model HAMSoC platform is illustrated in Figure 6.5, which consists of three parts: the formal specification of hierarchical agents and resources, the formal specification of monitoring operation, and the formal modeling of state transition of agents and resources.

The formal specification of agents and resources specifies the parameters of agents and resources, which are visible in the monitoring operations. System-level designers need to consider which parameters from agents or resources are needed

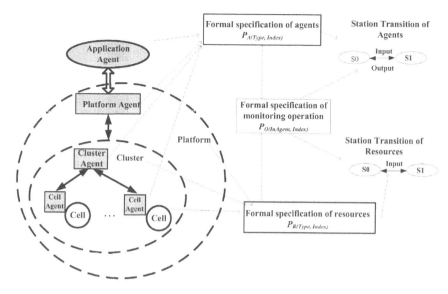

FIGURE 6.5
Formal specification framework of HAMSoC.

in the intended monitoring operations. Low-level designers need to consider how to realize the visibility of these parameters, if feasible. For example, when a system requires localized thermal management, the temperature of each chip region becomes a parameter abstracted on the system level. The low-level design will determine how the thermal sensors should be embedded.

The formal specification of a monitoring operation specifies all required information of the operation, including the initiating agents, the triggering condition of operations, the target resources, the function of the monitoring operation. These information must be visible from the parameters given in the specification of agents and resources, or from other monitoring operations.

Monitoring operations lead to state transition of resources and agents, which can also be triggered by implicit events, such as errors or physical changes. Any visible state transitions can be modeled by the updates of parameters specified to represent the involved agents or resources. The modeling of state transition enables in-depth analysis of system function and performance. The occurrence and duration of particular states influence the function and performance of related components and the whole system. For instance, the transitional period of voltage switching in SoC platform may prevent normal operation.

Table 6.2 summarizes the abbreviations, symbols and notations used for the formal specification.

TABLE 6.2

List of abbreviations and symbols

Symbol	Interpretation
AAG	Application agent
$PLAG$	Platform agent
$CLAG$	Cluster agent
$CEAG$	Cell agent
$()$	tuple
$\{\}$	set
$\emptyset$	empty set
$R(Type, Index)$	a resource uniquely identified by its Type and Index
$A(Type, Index)$	an agent uniquely identified by its Type and Index
$Type_{Index}$	the shorthand notation of a resource or agent, e.g. $Cell_i$
$P_{R(Type, Index)}$	the specification of a resource
$P_{A(Type, Index)}$	the specification of an agent
MP	the set of monitoring operations (issued by an agent)
$O(InAgent, Index)$	a monitoring operation uniquely identified by its initiating agent and Index
$P_{O(InAgent, Index)}$	the specification of an operation
$F_{R(Type, Index)}$	the FSM of a resource
$F_{A(Type, Index)}$	the FSM of an agent
$S_{R(Type, Index)}$	a state of a resource
$S_{A(Type, Index)}$	a state of an agent
$S_{R(Type, Index)\{V_i\}}$	a state of a resource with a variable V_i
$S_{A(Type, Index)\{V_i\}}$	a state of an agent with a variable V_i
$V_i \equiv v0$	V_i has a stable value of $v0$
$V_i \doteq (v0, v1)$	V_i is in transitional period from $v0$ to $v1$
$V_i \rightarrow v0$	an explicit input changing the value of V_i to $v0$
$V_i \dashrightarrow v0$	an implicit input changing the value of V_i to $v0$
$\leftarrow$	static assignment of values
$\Leftarrow$	dynamic assignment of values
$:$	a symbol to obtain an element from the specification of a resource or agent (a tuple)
$/$	a symbol connecting alternative contents in one expression
$[\,]$	boundary of alternative contents in one expression
$\lfloor\,\rfloor$	boundary of condition expression in the specification of monitoring operations
X	indicating an unused element in the specification of a monitoring operation
$=_{def}$	defined as
$S_{R(Type, Index)} \Rightarrow \Sigma \Rightarrow S'_{R(Type, Index)}$	the state of a resource transits from S to S' triggered by an input
$i\%n$	modulo operation
$\vee$	logical OR
$=$	comparison operator, equal to
$>$	comparison operator, larger than

6.3.2   Specification of agents and resources

Each agent or resource is specified as a tuple listing all parameters considered in the monitoring operations. By specifying these parameters, the designers determine the intended SW/HW interfaces.

6.3.2.1   Formal specification of resources

Definition 1 Let $R(Type, Index)$ represents a resource, with its type and index uniquely identifying the resource, then

$$P_{R(Type,Index)} =_{def} (Type, Index, Affi, P_{STA}, P_{OBS}, P_{REC})$$

is the specification of the resource.   □

In Definition 1, each element is explained as follows:

- *Type*: The type of resource. This element has enumeration value, for instance *Cell*, *Cluster* or *Platform*.

- *Index*: The index of the resource of all same-type resources. Any resource can be uniquely identified with *Type* and *Index*, thus a shorthand notation of a resource is $Type_{Index}$, for instance $Cell_i$ (the ith cell).

- *Affi*: The agent that this resource is affiliated to. This element specifies the position of the resource in the monitoring hierarchy. By specifying this element of all resources and agents, the whole monitoring hierarchy can be specified accurately, as illustrated in Figure6.6. Some examples of using *Affi* element:

  $Cell_i : Affi \hookleftarrow CEAG_i$

  $CEAG_i : Affi \nLeftarrow CLAG_j$

  where ": " is the symbol to obtain an element from the specification of a resource or agent (Table 6.2). Thus $Cell_i : Affi$ means the *Affi* of the resource $Cell_i$. As the lowest level of monitoring unit, the mapping of one cell agent to one cell is statically assigned. Thus static assignment $\hookleftarrow$ is used in the expression. A cell agent is dynamically configured into a cluster, thus its affiliation is dynamically assigned (denoted by $\nLeftarrow$) with $CLAG_j$ (the cluster agent). If a cell agent is not assigned to any cluster agent (a non-used cell), its affiliation is *NonAssigned*. It is important to differentiate between static and dynamic assignment, since dynamic assignment requires special low-level support to enable the run-time reconfiguration.

- P_{STA}, P_{OBS}, P_{REC}: Three types of parameters with different monitoring features. P_{STA} is the set of all static parameters of the resource. Static parameters are configured before execution, so no dynamic monitoring can change their values. But their values can be used in the monitoring processing. P_{OBS} is the set of all observable parameters, whose values can be observed at the run-time. This implicitly demands the support of low-level hardware. P_{REC} is the set of all reconfigurable parameters, whose values can be reconfigured at the run-time.

As a simple example for the three types of parameters, on a NoC platform, the voltage of all network links (each link is set as a cell) is fixed at 1.5V:

$$Cell_{i=1..N} : P_{STA} \leftarrow \{ \ Voltage \leftarrow 1.5V \ \}$$

The current driving the whole network (the platform) can be detected by a current detector:

$$Platform : P_{OBS} \leftarrow \{ \ Current \ \}$$

The frequency of each link can be reconfigured separately (with individual PLLs for instance):

$$Cell_{i=1..N} : P_{REC} \leftarrow \{ \ Frequency \ \}$$

In a multi-level hierarchy, a monitoring relation can cross more than one level; thus Recursive Affiliation is defined:

Definition 2 Let $R(Type, Index)$, and $A(Type, Index)$ be one resource and one agent, respectively, then

$$[A(Type, Index)/R(Type, Index)] : Affi_{|k, k>1|} =_{def}$$
$$[A(Type, Index)/R(Type, Index)] : Affi_{|k-1|} : Affi$$

$$[A(Type, Index)/R(Type, Index)] : Affi_{|1|} =_{def}$$
$$[A(Type, Index)/R(Type, Index)] : Affi$$

□

For instance, one cell with index i is statically affiliated to its cell agent, which is dynamically affiliated to a cluster agent with index j:

$$Cell_i : Affi_{|2|} \nleftarrow CLAG_j$$

Based on the recursive affiliation definition, we define Remote Affiliate and Direct Affiliate in Definition 3.

Affiliated to

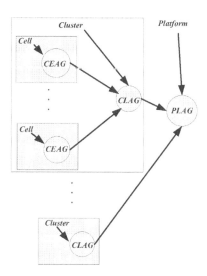

FIGURE 6.6
Illustration of affiliation relation in HAMSoC platforms.

Definition 3 Let $R(Type, Index)$, and $A(Type, Index)$ be one resource and one agent, respectively, then if

$$[A(Type, Index)/R(Type, Index)] : Affi_{|k,k>1|}[\hookleftarrow / \nleftrightarrow]A(Type', Index')$$

we say that $A(Type, Index)$ or $R(Type, Index)$ is remotely affiliated to $A(Type', Index')$; if

$$[A(Type, Index)/R(Type, Index)] : Affi_{|1|}[\hookleftarrow / \nleftrightarrow]A(Type', Index')$$

we say that $A(Type, Index)$ or $R(Type, Index)$ is directly affiliated to $A(Type', Index')$
□

6.3.2.2 Formal specification of agents

Similar to a resource, an agent can be formally specified as a tuple:

Definition 4 Let $A(Type, Index)$ represents an agent, with its type and index uniquely identifying the agent, then

$$P_{A(Type, Index)} =_{def} (Type, Index, Affi, P_{STA}, P_{OBS}, P_{REC}, MP)$$

which is the specification of the agent.
□

In Definition 4, *MP* is the set of all monitoring operations (Section 6.3.3) to be issued by the agent. The interpretations of all other elements are the same as them in the specification of resources. It should be noted that as we separate all adaptive monitoring functions into the abstraction of agents, only an agent is able to issue a monitoring operation.

6.3.3   Specification of monitoring operations

Monitoring operations are issued by agents, and targeted at resources or agents monitored by the initiating agents. The operations are triggered by certain conditions. We will first present the general format for specifying monitoring operations, and then introduce six types of operations defined on the HAMSoC system.

6.3.3.1   Specification format

Definition 5 A monitoring operation, denoted as $O(InAgent, Index)$ (the monitoring operation with a particular index of all operations issued by the initiating agent), can be specified as a tuple:

$$P_{O(InAgent, Index)} =_{def} (Type, InAgent, Index, Condition, Message,$$

$$[R(Type', Index')/A(Type', Index')]/ : [Affi/P_{OBS}/P_{REC}], New Value)$$

$\square$

In Definition 5, each element is defined as follows:

- *Type*: The type of the monitoring operation, enumeration value. We will explain the type in Section 6.3.3.2.

- *InAgent*: The initiating agent of the monitoring operation. Only an agent can initiate or issue a monitoring operation.

- *Index*: The index of the operation of all operations issued by the initiating agent. $O(InAgent, Index)$ uniquely identifies the operation.

- *Condition*: The condition of issuing a monitoring operation. The triggering condition of a monitoring operation can be specified using any defined parameter that is visible to the initiating agent, including the following three types:

  1. Clock that is local to the initiating agent.

  2. The evaluation of a visible parameter. A defined parameter is visible to an agent if the value of the parameter can be observed by an operation. For example, when the current of a cluster can be detected by a current sensor, the cluster agent can observe the current by issuing a *Trace* operation. An operation can be triggered when the value of a visible parameter meets a certain criteria, e.g. $\lfloor Cluster_i : Current > 1mA \rfloor$ (the current of the cluster is greater than $1mA$).

3. A monitoring operation: a monitoring operation visible to an agent can also trigger another monitoring operation. A visible operation to an agent is any operation that is issued by the same agent, or goes across the agent before reaching its target (e.g. the *Order* operation is used to monitor a remotely affiliated agent or resource).

The condition of a monitoring operation can be specified by any expression based on the previously mentioned types of parameters, for example

$$\lfloor StartTime, \lfloor N, Clk_{A(Type,Index)} \rfloor \rfloor$$

which defines a periodically issued monitoring operation, with the start time *StartTime* and the interval $\lfloor N, Clk_{A(Type,Index)} \rfloor$ (N cycles of the clock of an agent; $Clk_{A(Type,Index)}$ is a shorthand notation for $A(Type,Index) : Clk$).

- *Message*: The information sent in *Communicate* operation. We will explain it more in the description of *Communicate* monitoring operation.

- $[R(Type',Index')/A(Type',Index')]/ : [Affi/P_{OBS}/P_{REC}]$: This element specifies the parameter to be traced or reconfigured in the monitoring operation.

- *NewValue*: The new value to be assigned to the reconfigurable parameter or affiliation as specified in the previous element.

6.3.3.2   Types of monitoring operation

A type of monitoring operation represents one group of operations sharing a similar purpose with comparable protocols. Based on the hierarchical agent monitored system platform, we propose six types of monitoring operations illustrated in Figure 6.7, which cover a variety of potential operations fulfilling hierarchical monitoring operations.

The functions of each type of operation are as follows:

- *Communicate*: This type of operation is used to exchange information between application and platform agent. The interactions between the application agent and the platform agent are abstract and flexible messages, since the application is not aware of the platform details. The platform agent will decide how to respond to the messages from the application agent. This is different from the monitoring operations within the platform where initiating agents dictate to the resources or agents monitored. For *Communicate* operation, a *Message* element is provided.

- *Trace*: This type of operation is used by an agent to trace the run-time value of an observable parameter (in the P_{OBS} list) of a directly affiliated agent or resource (Definition 3).

- *Inquire*: This type of operation is issued by an agent to trace an observable parameter of a remotely affiliated resource or agent (Definition 3). For instance, a

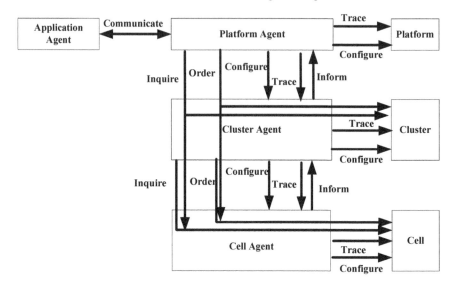

FIGURE 6.7
Types of monitoring operations.

platform agent can use this operation to check the current of a cluster. *Inquire* operation is different from *Trace* operation since it goes across multiple levels of monitoring hierarchy, and any *Inquire* operation needs to go through the lower level agent first before checking the targeted agent or resource (Figure 6.7).

- *Configure*: This type of operation is used by the initiating agent to modify a reconfigurable parameter of a directly affiliated agent or resource.

- *Order*: This type of operation is issued by an agent to modify a reconfigurable parameter of a remotely affiliated agent or resource.

- *Inform*: Issued by one agent to initiate a report of the run-time value of an observable parameter of a directly affiliated resource or agent, to the upper level agent which the initiating agent is directly affiliated to.

All types of monitoring operations can be formally specified using the general specification format (Definition 5). Examples will be given in Section 6.4.

6.3.4　State transition of agents and resources

During the monitoring operations, the values of parameters in the specification of agents and resources are being updated as a result of explicit reconfiguration commands, data processing and physical environment change. The run-time values of these parameters define the states of the agent or resource. Here we adopt FSM-based state transition modeling for early-stage analysis of hierarchical agent monitored systems, focusing only on monitoring-related reconfiguration. It provides a high-level abstraction of state parameters (variables) which are of interests to system-level designers. FSM is a well-developed and easy-to-understand formal model. This early-stage state transition modeling can be transformed into generic formal languages for automated property verification in future development.

6.3.4.1 State transition of resources

The finite state machine of each resource $R(Type, Index)$ is specified as a 4-tuple:

Definition 6 Let $R(Type, Index)$ represents a resource, then its state machine is a 4-tuple:

$$F_{R(Type, Index)} =_{def} (\Sigma, \bar{S}_{R(Type, Index)}, S0_{R(Type, Index)}, \delta)$$

Σ is the set of inputs. $\bar{S}_{R(Type, Index)}$ is the set of all possible states (state space). $S0_{R(Type, Index)}$ is the initial state. δ is the state transition function. □

Each state in the state space $\bar{S}_{R(Type, Index)}$ is defined as follows:

Definition 7 Let V_i represents a parameter, called a variable, with an index of i in the state of a resource $R(Type, Index)$, then

$$S_{R(Type, Index)}\{V_0, V_1, \cdots, V_{n-1}\}$$

represents the state of the resource $R(Type, Index)$. V_0, V_1, ..., to V_{n-1} are the n variables in its state. Each V_i represents a pair:

$$V_i =_{def} \{na_i, v_i\}$$

where na_i stands for the name of the parameter, and v_i stands for its value. When we write

$$S_{R(Type, Index)}\{V_i\}$$

it means that only the variable V_i is considered in a particular state. □

The state variables can be any parameters as defined in the specification of the agent or resource. As required for finite state machine, in case a variable has a continuous value range, for instance the physical parameters such as current or temperature, we discretize the value range into a finite number of values. For instance, the current can have the following values:

$$\{v0 =_{def} v \le 1A, \quad v1 =_{def} v > 1A\}$$

Then we define an expression: a variable V_i is stable with value $v0$ (Definition8).

Definition 8 Let V_i represents a variable in the state of a resource $R(Type, Index)$, then

$$S_{R(Type, Index)}\{V_i \equiv v0\}$$

means the variable has a stable value of $v0$. □

Definition 8 tries to model a situation where the values of a parameter is stable at a particular time, as seen by the monitoring operations. It is intended to differentiate the situation when a parameter is under reconfiguration, whose value is in transitional period. For instance, the voltage of a component may go through a noticeable transitional period before stabilizing, and its value is unstable during the period. The transitional value of a state variable is defined by Definition 9.

Definition 9 Let V_i represents a variable in the state of a resource $R(Type, Index)$, then

$$S_{R(Type, Index)}\{V_i \doteq (v0, v1)\}$$

means that the state variable V_i is in transitional period, whose value transits from $v0$ to $v1$.   □

There are two types of inputs to the FSM of a resource. One type is Explicit Input (Definition 10), which is a monitoring command from an agent.

Definition 10 Let V_i represents a variable in the state of a resource $R(Type, Index)$, then

$$V_i \rightarrow v0$$

represents an explicit input, which is an operation issued by an agent that reconfigures the value of the variable to $v0$.   □

The other type of inputs is Implicit Input, as a result of data processing or physical environment change that is transparent to the monitoring operations. For instance, as the system operates, the temperature may go implicitly from below 100 degrees to over 100 degrees, which goes to another value (note that the value of temperature is also discretized into value ranges). Such implicit input is defined by Definition 11:

Definition 11 Let V_i represents a variable in the state of a resource $R(Type, Index)$, then

$$V_i \dashrightarrow v0$$

represent an implicit input, which immediately triggers the value of the variable to switch to a value $v0$.   □

In fact, the switching of values as a result of implicit input does not happen immediately in the hardware. But such switching happens implicitly during data processing and communication or under physical environment influence, and the time it actually happens is not determined by the agents. Whenever the agents detect such change, it already has happened. Thus as seen by the agents, the input seems to have effect immediately.

Definition 12 defines a state transition, which exemplifies one variable being modified.

Definition 12 Let $S_{R(Type, Index)} \{V_0, V_1, \cdots, V_{n-1}\}$ represents the state of a resource, the

$$S_{R(Type, Index)}\{[V_i \equiv v0]/[V_i \doteq (v0, v1)]\} \Rightarrow V_i[\rightarrow / \dashrightarrow]v2 \Rightarrow$$

$$\{[V_i \equiv v2]/[V_i \doteq (v0, v2)]\}$$

represents a state transition of a variable V_i. The variable can be in either stable or transitional state before and after the transition. □

With the definitions of transitional/stable variables and explicit/implicit inputs, we can identify several types of state transitions common in adaptive reconfiguration process. With implicit input, a state variable, either stable or transitional, can transit to a stable state:

$$S_{R(Type, Index)}\{[V_i \equiv v0]/[V_i \doteq (v0, v1)]\} \Rightarrow V_i \dashrightarrow v1 \Rightarrow \{V_i \equiv v1\}$$

For instance, if the voltage of component is reconfigured to increase and enters the transitional period, whenever the voltage transition implicitly finishes (dependent on the voltage charging on power lines), the voltage enters the stable state with a new value.

With explicit input, a state variable, either with stable or transitional value, will transit to a transitional state:

$$S_{R(Type, Index)}\{[V_i \equiv v0]/[V_i \doteq (v0, v1)]\} \Rightarrow V_i \to v2 \Rightarrow \{V_i \doteq (v0, v2)\}$$

The consequence of transitional state after explicit input is due to the fact that most reconfiguration operations, as triggered by explicit input, take certain time to finish. Thus the variable always firstly enters the transitional state until the reconfiguration finishes (an implicit input), then becomes stable. However, if the reconfigurable period is so short as beyond the notice of the agents (for instance, within one cycle in a synchronous system), the transitional period may be omitted.

6.3.4.2 State transition of agents

The behaviors of agents are different from those of resources. First, it issues monitoring operations as outputs. Second, agents usually have fewer observable or reconfigurable parameters than resources, since the reconfiguration is mostly targeted at resources. The agents have one important observable parameter, *Error* (the error state). If the agent is error-free, it will provide the intended monitoring operations. Otherwise, it stops issuing correct monitoring operations. We can formally model the FSM of agents as a mealy machine (Definition 13).

Definition 13 Let $A(Type, Index)$ represents a resource, then its state machine is a 6-tuple:

$$F_{A(Type, Index)} =_{def} (\Sigma, \Gamma, \bar{S}_{A(Type, Index)}, SO_{A(Type, Index)}, \delta, \omega)$$

Σ is the set of inputs. Γ is the set of outputs. $\bar{S}_{A(Type, Index)}$ is the set of all possible states (state space). $SO_{A(Type, Index)}$ is the initial state. δ is the state transition function. ω is the output function. □

- $\bar{S}_{A(Type, Index)}$: The state space of an agent, which is the set of all potential states. An agent state $S_{A(Type, Index)}$ can be defined similarly as the resource state in Definition 7.

TABLE 6.3
Output function of agent state machine

State	Input	Output
$S_{A(Type,Index)}$ $\{Error \equiv 0\}$	$O(InAgent, Index)$: Condition	$O(InAgent, Index)$
$S_{A(Type,Index)}$ $\{Error \equiv 1\}$	*Don't Care*	*None*

- Σ and δ: Generally speaking, the state transition of agents can be modeled in the same way as that of resources. However, with most reconfiguration targeted at resources, there are few state variables which are needed to separately model the agents. For instance, a temperature variable is usually defined in the state of a resource, even though an agent may also be physically located in the same thermal region. In this case, there is no point modeling the variable into the agent's state since no reconfiguration is issued to deal with the agent's temperature separately. Nonetheless, there is one important state variable, *Error*, whose value will transit as a consequence of error occurrence (implicit input).

- Γ and ω: The outputs of agents are the monitoring operations, triggered by the conditions as specified in each operation. When the state variable *Error* is negative (error-free), the agent outputs an operation when the specified condition is satisfied. If the *Error* variable is positive, there is no output (or useful outputs) from the state machine with any inputs (Table 6.3).

6.4   Design example: hierarchical power monitoring in HAMNoC

In this section, the formal specification and modeling of the HAMSoC design approach will be demonstrated by a design example on Network-on-Chip. As a scalable on-chip communication architecture, the size of NoC is increasing rapidly in recent years [26, 305]. The monitoring services optimizing various metrics are widely used in different NoC architectures, in order to provide power management, fault tolerance, dependability and other services under the influence of circuit and workload variations.

The design example exemplifies hierarchical power monitoring operations, since power consumption has become a major design constraint. Various power management and monitoring techniques are applied on different structural and architectural levels, with different overheads. Globalized operations use large amount of global information with typically heavy data processing, and they affect the general configuration of the network. For instance, [137] presents energy-aware application mapping on NoC platforms. Considering the amount of communication between application segments, the optimal mapping to available processing elements in NoC can be determined. Such mapping process requires the information from application and all processing elements, and the outcome influences the global configuration of the platform. Such power management operations suit well into the functions of the platform agent or cluster agent, dependent on the relative size of required resources. Localized power management techniques only need local information with minimal amount of data processing. These operations typically change the configuration of local circuits or components. They can be performed much more frequently than globalized ones. For example, [158] presents adaptive link buffering in on-chip channels. The repeaters on the links can be dynamically configured as buffers under heavy network load. Such distributed and localized power optimization techniques require minimal information (network traffic in this case) and configuration time, and each link can be configured individually. Localized power management operations are suitable functions for the cell agent or cluster agent.

The design example adopts a set of power monitoring techniques applied onto a generic NoC platform. The formal specification and modeling of the involved hierarchical operations will be presented. We will show how HAMSoC approach captures the HW/SW interfaces with a high abstraction by adopting the previously described formal framework.

6.4.1 System description

The design example is given on a regular mesh-based NoC structure with hierarchical agent monitors (HAMNoC) illustrated in Figure 6.8. Each processing element with its corresponding switch and the four outgoing channels starting from the switch are configured as a cell, which is monitored by a cell agent. A number of cluster agents are distributed in the system. One possible manner of locating the cluster agent is to use a separate network node. Each of the cluster agent can be dynamically assigned with a number of cell agents by the platform agent. The platform agent and application agent are software modules centrally located in the system.

The set of monitoring operations exemplified in the design example is summarized in Table 6.4, which implies no restriction of the possible operations but serves as a demonstration of the design approach. The general aim of the monitoring services is to minimize the communication power. In particular, the system should run with as minimal voltage and frequency as possible, under an upper boundary of average latency. To achieve this, the application agent will inform the platform agent with the latency boundary as a requirement. The platform agent, after assigning clusters with cells, informs the cluster agents of the latency boundary. The cluster agent applies DVFS (dynamic voltage and frequency scaling) on each cluster. In detail, the cell agent will trace the average latency and received number of messages of each cell, and reports the values to the cluster agent. Each cluster agent then calculates the average latency of the cluster, and accordingly reconfigures the supply voltage, frequency and clock gating. Clock gating is applied during the voltage transition since it provides stable supply voltage during the transition [296].

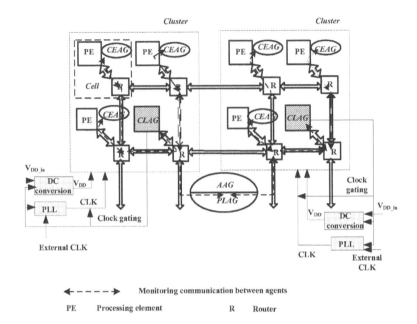

FIGURE 6.8
An example of an HAMNoC platform.

TABLE 6.4
Hierarchical power monitoring operations exemplified in the design example

Level of Agents	Monitoring Operations
Application Agent	determine the average latency boundary and inform the platform agent
Platform Agent	cluster assignment determine latency boundary of each cluster
Cluster Agent	monitor latency of each cluster apply DVFS and clock gating
Cell Agent	trace local traffic volume and latency report them to cluster agent

6.4.2 Specification of resources

The platform is specified as follows:

$$Platform_0 =_{def} (Platform, 0, PLAG, \emptyset, \{Latency\}, \emptyset\})$$

$$Platform_0 : Latency \nleftarrow \frac{\sum_{i=0..(n-1)} Cluster_i : Latency}{\sum_{i=0..(n-1)} Cluster_i : Received}$$

For the platform, there is one observable parameter *Latency* in the P_{OBS} list. It is the average latency of all messages transmitted in the network. The latency of the platform is calculated by the *Latency* (total latency of all received messages) of each cluster and the *Received* (the number of received messages) in each cluster. These two values are sent by cluster agents with *Inform* operation to the platform agent.

Each cluster (index $0, .., n - 1$) is specified as follows:

$$Cluster_{i=0..(n-1)} =_{def} (Cluster, 0..(n - 1), CLAG_{0..(n-1)}, \emptyset,$$
$$\{Latency, Received\}, \{Volt, Freq, ClkG\})$$

For each cluster, there are three reconfigurable parameters: *Volt* (voltage), *Freq* (frequency), and *ClkG* (clock gating). There are two observable parameters:*Latency* (the average latency of the communication in the cluster) and *Received* (the received number of messages). They can be calculated based on the information of each cell in the similar way as the platform latency is calculated, and the involved specification is omitted here.

Each cell (index $0, .., (m \times n - 1)$) is specified as follows:

$$Cell_{i=0..(m \times n-1)} =_{def} (Cell, 0..(m \times n - 1), CEAG_{0..(m \times n-1)}, \emptyset,$$
$$\{Latency, Received\}, \emptyset)$$

For each cell, there are two observable parameters: *Latency* (the total latency of all received messages), and *Received* (the total number of received messages). There are no reconfigurable or static parameters defined.

6.4.3 Specification of agents and monitoring operations

The application agent is specified as follows:

$$AAG_0 =_{def} (AAG, 0, X, \emptyset, \emptyset, \emptyset, MP)$$
$$AAG_0 : MP \nleftarrow \{(Communicate, AAG_0, 0, \lfloor Startup \rfloor, LatencyB, X, X)\}$$

There is one defined monitoring operation for the application agent: a *Communicate* operation, by which the application agent sends the latency boundary, *LatencyB*, to the platform agent.

The platform agent is specified as follows:

$$PLAG_0 =_{def} (PLAG, 0, X, \emptyset, \{Clk\}, \{LatencyB\}, MP)$$

$$PLAG_0 : MP \hookleftarrow \{$$

$$(Configure, PLAG_0, 0, \lfloor O(AAG_0, 0) \rfloor, X,$$

$$CLAG_{0..n-1} : LatencyB, PLAG_0 : LatencyB),$$

$$(Trace, PLAG_0, 1, \lfloor Startup, \lfloor W, Clk_{PLAG_0} \rfloor \rfloor, X,$$

$$Platform_0 : Latency, X)\}$$

The platform agent performs two operations. The first is a *Configure* operation which configures the *LatencyB* (latency boundary) of all clusters as the latency boundary of the platform. This operation is triggered when the platform agent receives the *Communicate* operation ($O(AAG_0, 0)$) from the application agent. The other operation is a periodical *Trace* operation which checks the latency of the platform every W cycles of the *PLAG* clock.

Each cluster agent is specified as follows:

$$CLAG_{i=0..n-1} =_{def} (CLAG, 0..(n-1), PLAG, \emptyset, \{Clk\}, \{LatencyB\}, MP)$$

$$CLAG_{i=0..n-1} : MP \hookleftarrow \{$$

$$(Configure, CLAG_i, 0, \lfloor Cluster_i : Latency > \rho \times LatencyB \rfloor,$$

$$X, Cluster_i : Volt, Hi(CLAG_i : Volt)),$$

$$(Configure, CLAG_i, 1, \lfloor Cluster_i : Latency > \rho \times LatencyB \rfloor,$$

$$X, Cluster_i : Freq, Hi(CLAG_i : Freq)),$$

$$(Configure, CLAG_i, 2, \lfloor Cluster_i : Latency \leq \rho \times LatencyB \rfloor,$$

$$X, Cluster_i : Volt, Lo(CLAG_i : Volt)),$$

$$(Configure, CLAG_i, 3, \lfloor Cluster_i : Latency \leq \rho \times LatencyB \rfloor,$$

$$X, Cluster_i : Freq, Lo(CLAG_i : Freq)),$$

$$(Configure, CLAG_i, 4, \lfloor O(CLAG_i, 0 \vee 1 \vee 2 \vee 3) \rfloor,$$

$$X, Cluster_i : ClkG, 1),$$

$$(Configure, CLAG_i, 5, \lfloor \lfloor O(CLAG_i, 4 \rfloor + T_d \rfloor,$$

$$X, Cluster_i : ClkG, 0),$$

$$(Inform, CLAG_i, 6, \lfloor Startup, \lfloor W, Clk_{CLAG_i} \rfloor \rfloor,$$

$$X, Cluster_i : Latency, X),$$

$$(Inform, CLAG_i, 7, \lfloor Startup, \lfloor W, Clk_{CLAG_i} \rfloor \rfloor,$$

$$X, Cluster_i : Received, X)\}$$

The cluster agent for each cluster performs eight operations (with the indexes 0–7). When the average latency of the cluster exceeds the latency boundary (multiplied by a coefficient ρ, $0 < \rho < 1$, to leave a margin to prevent temporary overshooting), both the voltage and frequency will be increased (operation 0 and 1). Otherwise if the average latency of the cluster is lower than the latency boundary, the voltage and frequency will be decreased (operation 2 and 3). $Hi(Volt/Freq)$ and $Lo(Volt/Freq)$ are two pre-defined functions returning the higher/lower level of voltage and frequency values, respectively. When the voltage and frequency are to be switched, clock gating is also enabled (operation 4) to avoid unstable communication during the transitional period. Thus the clock gating is issued when any of the operations indexed 0–3 is issued (the condition expression $\lfloor O(CLAG_i, 0 \vee 1 \vee 2 \vee 3) \rfloor$ indicates the logic OR of operations 0–3). After a delay of T_d, when the voltage transition finishes, the clock gating will be disabled to restart communication (operation 5). Operations indexed 6 and 7 inform the latency and received message number to the platform agent. Each cell agent is specified as follows:

$$CEAG_{i=0..(m \times n - 1)} =_{def} (CEAG, 0..(m \times n - 1), CLAG_{i\%n}, \emptyset, \{Clk\}, \emptyset, MP)$$

$$CEAG_{i=0..(m \times n - 1)} : MP \leftarrow \{$$
$$(Trace, CEAG_i, 0, \lfloor Startup, \lfloor W', Clk_{CEAG_i} \rfloor \rfloor,$$
$$X, Cell_i : Latency, X),$$
$$(Trace, CEAG_i, 1, \lfloor Startup, \lfloor W', Clk_{CEAG_i} \rfloor \rfloor,$$
$$X, Cell_i : Received, X),$$
$$(Inform, CEAG_i, 2, \lfloor \lfloor Startup, \lfloor W', Clk_{CEAG_i} \rfloor \rfloor + T_c \rfloor,$$
$$X, Cell_i : Latency, X),$$
$$(Inform, CEAG_i, 3, \lfloor \lfloor Startup, \lfloor W', Clk_{CEAG_i} \rfloor \rfloor + T_c \rfloor,$$
$$X, Cell_i : Received, X)\}$$

Each cell agent performs four operations. Periodically, each cell agent traces the latency and received message number of each cell (operation 0 and 1). Also periodically, the cell agent informs the cluster agent of the two values (operation 2 and 3). The *Inform* operations are delayed by a short time T_c (specified by low-level design) compared to the time of *Trace* operations, so that the values returned by *Trace* operations are ready to be sent. $i\%n$ represents a modulo function, since modulo mapping of cell indexes to cluster indexes is assumed.

It should be noted that the specification of basic initialization operations are omitted here, for instance affiliating cell agents to cluster agents and the initial configuration of cluster voltage, frequency and clock gating.

6.4.4   Formal modeling of state transitions

The state transition of agents in this design example is quite straightforward, as agents perform the monitoring operations based on the condition specified in the operations. Of all levels of resources, most configurations are performed on clusters. Thus we exemplify the specification of FSMs of clusters with detailed explanations.

Firstly we need to determine the state variables. The monitoring service performed on clusters is the reconfiguration of voltage, frequency and clock gating based on the latency values. Thus four variables are included in the state:

$$S_{R(Cluster,i=1..n-1)}\{Latency, Volt, Freq, ClkG\}$$

All these variables need to have a finite number of value alternatives, thus discretization is performed. Starting with latency:

$$\{l_h =_{def} Latency > \rho \times LatencyB, \quad l_l =_{def} Latency \le \rho \times LatencyB\}$$

When the latency is higher than the latency boundary (*LatencyB* is the latency boundary set by the application agent), it has a value l_h (high latency), otherwise the other value l_l (low latency). As the value of latency is always implicitly changed (as a result of network communication), there is no transitional state. Then voltage and frequency:

$$\{v_l, (v_l, v_h), v_h\}$$

$$\{f_l, (f_l, f_h), f_h\}$$

We assume there are two alternative supply voltages and frequencies available for switching, thus they each has one high value, one low value and one transitional value. For clock gating, it is either 0 (non-gated) or 1 (gated). As clock gating usually takes minimal delay, the transitional period can be omitted (if needed, the transitional value can be defined similarly to that of voltage or frequency as well).

With these state variables, we can model the state transitions of each cluster as in Figure 6.9.

For brevity, we omit writing the variable name with values listed in the order of {*Latency, Volt, Freq, ClkG*}. We explain some transitions in more detail as below:

$$S_{R(Cluster,i=0..n-1)}\{(f_l, f_h)\} \Rightarrow Freq \dashrightarrow f_h \Rightarrow \{f_h\}$$

The frequency transition implicitly finishes. This transition concerns only the variable *Freq* (regardless of the values of other variables).

$$S_{R(Cluster,i=0..n-1)}\{(v_l, v_h)\} \Rightarrow Volt \dashrightarrow v_h \Rightarrow \{v_h\}$$

The voltage transition implicitly finishes.

$$S_{R(Cluster,i=0..n-1)}\{ClkG \equiv 1\} \Rightarrow ClkG \to 0 \Rightarrow \{ClkG \equiv 0\}$$

The clock gating will be disabled by operation $O(CLAG_i, 5)$.

$$S_{R(Cluster,i=0..n-1)}\{l_h, v_h, f_h, 0\} \Rightarrow Latency \dashrightarrow l_l \Rightarrow \{l_l, v_h, f_h, 0\}$$

After the new voltage and frequency are applied, the latency may implicitly drop to low.

This design example mainly serves demonstrative purpose for the formal specification framework. To avoid lengthy scenario-specific description, certain assumptions are adopted. For instance, $W' > T_d$ (the interval of latency tracing is longer than the transitional time of voltage), so that the power reconfiguration always finishes before the following reconfiguration.

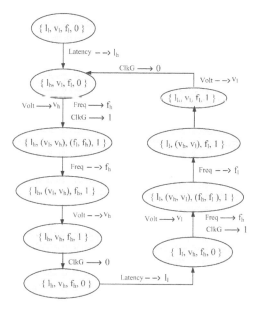

FIGURE 6.9
Exemplified state transition graph for cluster.

6.5   Conclusions

This chapter presented a novel design approach, hierarchical agent monitored SoC (HAMSoC), by elaborating its formal specification framework and demonstrating a design example of hierarchical power monitoring on NoCs.

HAMSoC follows monitoring-centric design, which focuses the design efforts on the realizing monitoring operations. It addresses the emerging challenges of dependability and variability in parallel or distributed embedded computing, from a novel design-method perspective. Particularly, on massively parallel SoC systems, various monitoring operations, based on their granularity and resource utilization, are mapped to the functions of agents on different system levels. Depending on the algorithm complexity and system constraints, agents are synthesized as hardware, software or hybrid. This approach provides scalability in terms of both design effort and physical overhead.

The formal specification framework of HAMSoC enables efficient system-level specification and modeling for early-stage development. It adopts a highly-abstracted formal language to specify exclusively the monitoring operations and parameters, with proper exposure of SW/HW interfaces. FSM model is used for state transition analysis in monitoring operations. The notations are clearly defined with practical interpretations on SoC system. The specification will be transformed into generic formal languages so that existing verification tools can be used for further system development, which is our current research work.

6.6   Glossary

Agent: The term has a variety of interpretations in different contexts. Here an agent refers to an abstract entity performing monitoring operations on certain resources. An agent can be implemented as a physically unified component (software, hardware, or hybrid), or as distributed components working together as a unified entity.

HAMSoC: Hierarchically agent monitored System-on-Chip, a design approach, following the general monitoring-centric methodology, tailored for massively parallel SoC systems with hierarchically embedded agents as monitors.

HAMNoC: Hierarchically agent monitored Network-on-Chip, a design approach, following the general monitoring-centric methodology, tailored for Network-on-Chip platforms with hierarchically embedded agents as monitors.

Monitoring-Centric Design: A novel design methodology proposed in this chapter to promote the design concentration on monitoring operations and interfaces for dependable and adaptive complex system design.

Monitoring Operation: An operation for monitoring functions, for instance power optimization, fault detection and recovery, performance tuning, etc., in contrast to computation and communication operations.

Resource: Resources, in monitoring-centric design, are the components under the monitoring of corresponding agents, regardless of their implementation.

Acknowledgment

The authors have been supported by the Foundation of Nokia Corporation.

7

Toward Self-Placing Applications on 2D and 3D NoCs

L. Petre, K. Sere, L. Tsiopoulos

Department of Information Technologies, Åbo Akademi University, Turku, Finland

P. Liljeberg, J. Plosila

Department of Information Technology, University of Turku, Turku, Finland

CONTENTS

7.1   Introduction

A single chip system consists of computing cores connected to memory elements and various application-specific cores, I/O cores, etc. Such chip systems are typically referred to as Systems-on-Chip (SoC) [105]. In recent years, the communication paradigms for the cores on the chip have evolved from bus-based and point-to-point solutions to Network-on-Chip (NoC) models [36]. The bus-based model grants access of only one master core at a time to the bus in order to establish communication with some other component. This limits the efficiency and the scalability of the whole system. In point-to-point solutions, every component can be connected directly to the other components, thus resulting in a large number of connections and increased complexity. The NoC communication paradigm scales better than the above mentioned paradigms as well as routes information in a distributed manner; hence, the NoC-based applications will run faster. With these advantages, it is likely that the NoC technology will become part of our everyday life infrastructures. In Figure 7.1,

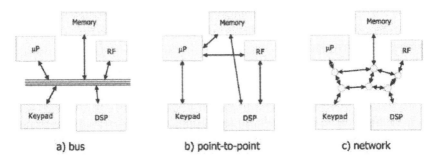

a) bus   b) point-to-point   c) network

FIGURE 7.1
Communication paradigms on a SoC.

originally published in [36], we show the evolution of the communication paradigms on a SoC and in Figure 7.2 we show a two-dimensional mesh model of a NoC. On a SoC employing the NoC communication paradigm, the various components of the system, referred to as *cores*, are connected among themselves using a network of switches and routers that exchange data packets, much as in a real network.

One of the recent technological approaches to designing chips has been a three-dimensional solution for a computing platform in which heterogeneous nodes such as processors, memory units, routers, etc are not placed anymore in a two-dimensional (2D) mesh only, but rather in a spacial configuration (3D) [222] employing the so-called 3D stacking technology where several silicon layers are stacked vertically [222]. The 3D technologies address the integration of many heterogeneous components much better, as each layer can support a different technology [233]. These cubic meshes are expected to display *massive parallelism*, where hundreds of processing nodes are running in parallel and a *distributed memory architecture*, so that storage is available close to the processing elements, with short access time. The NoC paradigm is very useful for 3D chips as it provides arbitrarily good scalability of interconnections and offers a high degree of parallel computation *and* parallel communication. Overall, due to extremely short distances between the stack layers, 3D NoCs have very good potential for both high performance and high energy efficiency. It is also apparent that the relative positioning of the various nodes influences performance since their communication can take longer if the nodes are physically further from each other. The vertical wires in a 3D structure are typically short and thick compared with the horizontal wires and are implemented using *Through Silicon Vias (TSV)* across layers. They can actually be very fast. However, the large physical diameter of a TSV indicates that the routers of a mesh cannot be equipped with full-width (highly parallel) vertical links but inter-layer transmission must be serialized through a single or few high-speed TSVs at each router.

These technological advances will achieve a great impact only if they are trustworthy, that is, if their functioning is guaranteed to provide their promises. *Formal methods* refer to one area of computer science which has already proved useful in the

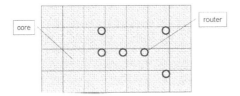

FIGURE 7.2
Two-dimensional NoC: cores are placed on the tiles (squares) and the routers (circles) handle the communication among cores.

verification of various hardware systems [32], via model checking, test generation, refinement, and theorem proving. A formal method provides essentially modeling based on mathematics. This implies the capture of system requirements in a specific, precise format. Importantly, such a format can be analyzed for various properties and, if the formal method permits, also stepwise developed (*refined*) until an implementation is formed. By following such a formal development, we are *sure* that the final result correctly implements the requirements of the system. Hence, formal methods can ensure the trustworthiness of various technologies.

Various drawbacks also need to be observed with respect to formal methods. One of them is their scalability problem, i.e., while it is feasible to model and verify critical parts of various systems it is not equally clear how to model and verify large systems. Tool support is another sensitive aspect especially associated with the difficulty of using formal methods as they are based on mathematics. Proving logical properties about systems is not obvious, in addition to the error-prone writing of specifications and proofs by hand. However, as shown by a growing body of literature, formal methods can be valuable when used in proper contexts.

With these observations we set out to propose a formal method-based language for placing various applications in an efficient manner on a 2D and on a 3D NoC. This language models heterogeneous, spatially stacked tiles containing computing units running in parallel. We aim toward enabling applications with the ability to discover a suitable placing strategy, so that the tiles with units communicating often are in a close proximity. Hence, this will become a quality of *self*-placing that applications residing on NoCs will display. We assume that the communication mechanism between the routers is based on message passing. Namely, when two modules communicate via shared data, the new values for the data are embedded in a message and transmitted between routers. When two modules communicate via procedure calls, the respective parameters are marshalled and unmarshalled using messages. However, message passing as such is outside the scope of this chapter.

We base our work on MIDAS [242], a language introduced for modeling middleware aspects for computing in real networks. The fundamental feature of MIDAS is the proposal of a separation of concerns with respect to functional and non-functional aspects. The functioning of networked applications is specified first and then, on top of this model, aspects of non-functional properties are superposed [150]. In [242] we show how to superpose the availability of a network for supporting a network application whose model is network generic; hence, we *adapt* the application to a specific network. Superposition is modeled with superposition refinement [13]. This essentially means that, by making an application aware of the network availability of its supporting network, code will be enabled less often; in addition, more code is added, not affecting the functional code but instead taking care of the availability issues, e.g., replication or node recovery.

When employing MIDAS for applications running on NoCs, we essentially take the same approach. We first model the functional aspects of an unplaced application and then we uncover a mapping of modules to NoC tiles, so that efficiency is supported by the proposed mapping. We then add more (monitoring) code to take care of the dynamic tile availability to the application and address the runtime replacing of the application.

One important feature that our approach promotes is the high-level of abstraction provided by the proposed modeling. We can gradually increase the level of detail when specifying the unplaced application using the *refinement* approach. Refinement techniques ensure that a high-level system specification can be transformed by a sequence of correctness preserving steps into an executable and more efficient system that satisfies the original specification. More importantly, we can then model various features onto this unplaced application toward placing it, such as movement and replication and we are still at a conveniently high-level of abstraction so that the real implementation details can be achieved in many ways. Thus, our refinement-based approach provides for more flexibility in placing applications on NoCs than other existing solutions.

We proceed as follows. In Section 7.2 we discuss related approaches and in Section 7.3 we outline our proposed adaptation of MIDAS to model applications running on NoCs. In Section 7.4, we introduce our placing algorithm and replacing techniques and emphasize their significance. In Section 7.5 we conclude the chapter.

7.2   Related work

Many approaches already exist on the topic of application mapping to NoC platforms. This mapping can be either *static*, when the application is first analyzed and then statically mapped to a NoC platform or *dynamic*, when the application mapping can be reconfigured at runtime, for instance, based on various observations from the application execution. Furthermore, the dynamic application mapping approaches can be classified as *centralized* – if the remapping of the application is handled by a central managing unit or *decentralized* – if the remapping is handled in a distributed manner. In this section we briefly review some of the existing mapping approaches. Based on this overview we then motivate our proposed method.

Hu and Marculescu [133] are the first to have proposed an approach to the mapping problem. Their branch and bound algorithm maps statically a given set of IP cores onto a regular NoC architecture. The total communication energy is minimized and the performance of the resulting communication system is guaranteed to satisfy the design constraints through bandwidth reservation. Chou and Marculescu present a contention-aware mapping approach for NoCs based on deterministic routing schemes with the goal of minimizing the end-to-end data packet latency in the network [63]. Their method first evaluates the impact of three different types of contention, namely source-based, destination-based and path-based contention. Then, a linear programming-based mapping technique is applied statically, aiming at minimizing the network contention and the energy consumption of the communication.

Addo-Quaye [4] presents an approach for thermal and communication-aware mapping and placement for 3D NoC architectures. Genetic algorithms are used in order to model static solutions to the mapping problems. The approach is shown to reduce communication and peak temperature when compared with purely random placements. Murali et al. [222] present a static method for establishing a power-efficient topology of a 3D System-on-Chip for a given application and for finding paths for the traffic flows that meet various Through Silicon Vias constraints. The method accounts for power and delay of both switches and links. The assignment of resources to different 3D layers and the floorplan of the resources in each layer are taken as inputs to the synthesis process, while the output of the process is the optimal positions of NoC switches in each layer and between the layers.

Chen et al. [58] propose a compiler framework with four major steps for a static energy-efficient application mapping to NoC-based Chip Multi-Processors (CMP). The first step parallelizes the application code and maps the resulting parallel threads to virtual processors. The second step implements a virtual-to-physical processor mapping. The third step maps data elements to the memories attached to the CMP nodes in order to place a data item into a node close to the nodes that access it. The fourth step determines the paths between memories and processors for data to travel in the NoC in an energy-efficient manner.

Fekete et al. [96] present an approach for simultaneous scheduling and placement of communicating modules for SoC architectures including devices with partial reconfiguration support. This makes their approach one of the first that addresses dynamic aspects for the mapping. Accurate models of reconfigurable hardware such as FPGAs can be created including reconfiguration times and space management of free and occupied resources. An optimization model is then used in an approach employing a linear programming solver to statically compute optimal hardware-software partitions with respect to execution times.

Yang et al. [327] propose a mapping strategy for block cipher security algorithms decomposed into tasks. Each task is mapped to a processing element of a NoC platform. The proposed approach is simulated by a cycle-accurate SystemC model platform called Networked Processor Array. The algorithms are written in C and are then profiled to identify execution groups which can be performed concurrently or sequentially. The scheduling and mapping step has as inputs the results from the profiling step as well as the available number of processing elements on the NoC platform. This strategy thus approaches a dynamic mapping by first simulating the system behavior and then map the system to the platform accordingly.

Smit et al. [278] present one of the first proper approaches for dynamic application mapping to heterogeneous and reconfigurable SoC architectures. Their algorithm determines the weight of a minimum processor assignment for any weighted process graph to a set of processors. Various constraints important for the efficient mapping of processor cores to tiles are considered in the algorithm, such as the processor capacity (e.g., at most one process on one processor) and the availability and the uniqueness of processors. The algorithm was further improved by changing the order of the task assignments to processors when considering the scarcity of resources and the processing costs of the processes.

Link and Vijaykrishnan [186] propose the use of dynamic runtime reconfiguration to shift hot spot-inducing computations on NoCs in order to make the thermal profile of the system more uniform. The transformation of the existing configuration information at runtime is performed by a migration unit which includes three migration functions: rotation, mirroring and translation shifting. Congestion-free operation is ensured by transforming groups of processing elements in phases.

Berthelot et al. present a system-level design flow methodology to address the mapping and scheduling of application tasks onto heterogeneous and reconfigurable architectures [34]. The design flow incorporates graphs to represent both the architecture and the application and aims at obtaining a near optimal scheduling and mapping of the application tasks onto a heterogeneous architecture. The implementation of the dynamic reconfiguration on hardware components is automated by a tool named SynDEx that also achieves an automated hardware/software mapping and scheduling.

Chou and Marculescu [64] propose a dynamic strategy for allocating application tasks to platform resources in NoCs. A master processor incorporates the user behavior information in the resource allocation process, which allows the system to better respond to real-time changes and adapt dynamically to user needs. Chou et al. [65] propose incremental mapping of dynamically incoming applications to NoCs with processing elements operating at multiple voltage levels. A global manager handles the system resources incorporating a dynamic mapping technique which allocates the appropriate resources to incoming application tasks. The existent system configuration is not affected by the incremental mapping process and the total communication energy consumption is minimized. Ding et al. [80] propose a centralized application mapping technique on CMPs that takes into account process variations and focuses on how to adapt the application execution at runtime to the variations. To further improve the efficiency of the technique, changing of the cores' frequencies is studied.

Kandemir et al. [148] present dynamic thread and data mappings for CMP distributed on NoC communication architectures with the goal to reduce the distance between the cores requesting data and the cores whose local memory contain the requested data. Helper threads which run in parallel with the application threads carry out the tasks of thread-to-core and data-to-core assignments. The integrated and dynamic thread-data mapping scheme achieves substantial savings in execution latency over static mappings.

Streichert et al. [288] propose an approach for topology-aware placement of replicated application tasks running on embedded networks. A two-phase and decentralized methodology addresses the integration of new tasks into the network and treats resource defects. A fast repair phase activates replicas and reroutes the communication paths. An optimization phase improves the binding of the tasks and places replicas in order to tolerate further resource defects.

Al Faruque et al. [95] present the first scheme for runtime application mapping in a distributed manner using agents. Virtual clusters are constructed at runtime and distributed application mapping within each cluster is accomplished by agents that are autonomous, modifiable and exhibit adaptation capabilities. The overall monitoring traffic produced to collect the current state of the system and needed for runtime mapping is vastly reduced in comparison to centralized mapping schemes. It also requires less execution cycles when compared with non-clustered centralized approaches.

Zipf et al. [337] propose a dynamic and decentralized task mapping approach for homogeneous multiprocessor NoCs. The approach employs the incremental application of three task mapping algorithms: an *exact* algorithm applied initially only on small modules, a *constructive* algorithm which produces a feasible initial solution and an *improvement* algorithm based on the constructive algorithm that is executed on every node of the network until a steady state is reached. A model of *spring-connected weights* is used as the basis of the improvement algorithm wherein weights correspond to tasks and springs to communication between the tasks.

Nickschas and Brinkschulte [227] introduce a real-time middleware using an auction-inspired approach for autonomous and decentralized task allocation. The middleware employs a service-oriented architecture where services are implemented as intelligent agents that make decisions related to their autonomic management: *self-configuration, self-optimization,* and *self-healing.* The agents cooperate using an auctioning mechanism based on Contract Net. They evaluate whether a given job can be carried out at all, using only local data and if so, they also evaluate the qualities that can be achieved and the costs that result from the processing.

In dynamic and centralized application mapping techniques issues such as the single point of failure (for the central manager) and the large volume of monitoring traffic may appear. Furthermore, centralized approaches may not be able to provide the required scalability for future 3D NoCs, thus, decentralized approaches are more suitable.

In this chapter we present several types of placing techniques, starting from a static and centralized approach used initially at design time and then exploring a distributed dynamic placing method as well as centralized dynamic approaches. The main difference with the aforementioned approaches is that we employ a formal specification language and a middleware based on that language in order to provide efficient application mapping. To the best of our knowledge, this is the first application mapping approach to employ a formal specification language. Our contribution is thus twofold. First, we adapt a middleware language initially dedicated to real networking to model applications running on 2D and 3D NoC meshes. Second, we define several approaches toward mapping such applications on 2D and 3D tile meshes, underlying various features enabled by the base formal method language.

7.3 NoC-oriented MIDAS

In this section we describe an adaptation of MIDAS to modeling applications running on NoCs.

The application is first specified via its modules that communicate with shared data only, for simplicity. Each module has a *location* (NoC tile) and the locations can form a two- or three-dimensional mesh. We use corresponding two- or three-dimensional matrices to denote these meshes, named $Tiles_{m \times n}$ and $Tiles_{m \times n \times p}$ respectively, where m, n, p stand for the number of elements on each dimension. An element of $Tiles_{m \times n}$ is denoted $Tiles(i, j)$, where $1 \leq i \leq m$ and $1 \leq j \leq n$ and an element of $Tiles_{m \times n \times p}$ is denoted $Tiles(i, j, k)$, where $1 \leq i \leq m$, $1 \leq j \leq n$, and $1 \leq k \leq p$. When we are not interested in the number of the mesh dimensions, we refer to the finite set of locations (tiles) as *Tiles*.

The modules can be either memory units, code resources, or even more complex processing elements. All the modules can run in parallel, independently of their locations. Communication of tiles is assumed fastest via the direct neighbors in the mesh; this is a simplifying assumption at this abstraction level, since the vertical links in a 3D NoC are very different from the horizontal ones.

In the following we first describe the memory and code resources and then the processing elements that perform the actual computing.

Memory units A *memory unit* (or a *data unit*) is modeled by a variable in our approach. Let *Memory* be a finite set of memory unit *names*. Memory units are specified as quadruples (mu, loc, Val, val) where $mu \in Memory$ is the name of the memory unit, $loc \in Tiles$ is its location, *Val* is a nonempty set of *values* denoting the memory unit type, and $val \in Val$ is the current value of the memory unit. We denote the location of a memory unit *mu* with the expression *mu.loc* and the names of the memory units located at a location $\alpha \in Tiles$ with the set $\alpha.memoryunit$. The value of a memory unit *mu* is given by the expression *mu.val*. We assume that the type of a memory unit is unchangeable. When $|mu.loc| > 1$ we say that the memory unit is *replicated*.

Code resources A *code resource* (or a *code unit*) is modeled by an action in our approach [242]. Let *Code* be a finite set of code resource *names*, distinct from *Memory*. Code resources are specified as triples (c, loc, C) where $c \in Code$ is the name of the action, $loc \in Tiles$ is its location, and *C* is its body, i.e., a statement that can model evaluation and updates of the memory units. We denote the location of a code resource (c, loc, C) with the expression *c.loc* and the bodies of the code resources located at a location $\alpha \in Tiles$ with the set $\alpha.coderesource$. We assume that the name of a code resource is unchangeable. When $|c.loc| > 1$ we say that the code resource is *replicated*. The body *C* of a code resource named *c* can model deadlocking and stuttering code, assignments, guarded code, sequential, conditional, and non-deterministic composition of a finite number of code resources. Semantically, the code resources are defined using *weakest precondition predicate transformers* [13]. The body $C = g \rightarrow B$ of the code resource *c* can execute only if the specific boolean condition g, named *guard condition*, holds; in this case we say that the code resource is *enabled*. Guards are formally defined using wp predicate transformers [13].

Processing elements A processing element is specified as

$$
\mathcal{PE} = \|[\quad
\begin{array}{ll}
\textbf{exp} & exp\_mu; \\
\textbf{var} & local\_mu; \\
\textbf{imp} & imp\_mu; \\
\textbf{do} \; \|_{i \in I} \, Code_i \; \textbf{od} &
\end{array}
\quad]\|
\tag{7.1}
$$

The first three sections are for memory unit declaration and initialization, while the last describes the computation involved in $\mathcal{PE}$, when I is finite. We assume that exp_mu, $local\_mu$ and imp_mu are sets of memory units whose names are pairwise disjoint, i.e., the name of a memory unit is unique in a processing element. This is a well-definedness constraint of the base action system framework [13].

The **exp** section describes the finite set of *exported* memory units exp_mu of $\mathcal{PE}$. These memory units are defined and can be used within $\mathcal{PE}$, as well as within other elements that import them. As the memory units can be imported by other elements, their names are unchangeable. The **var** section describes the finite set of *local* memory units $local\_mu$ of $\mathcal{PE}$. These memory units are defined and can be used only within $\mathcal{PE}$. As the memory units are local to $\mathcal{PE}$, their names can be changed. This change has to respect the requirement of unique names for memory units in a processing element and has to be propagated in all the code resource bodies that use the respective local memory units. The **imp** section describes the finite set of *imported* memory units imp_mu. These memory units are specified by name and possibly desired locations of import, denoted $imp\_mu_k.iloc$, where $imp\_mu = \{imp\_mu_k\}_{k \in K}$ and K is a finite index set. The imported memory units are used in $\mathcal{PE}$ and are declared as exported in other processing elements. The imported and the exported memory units form the *global* memory units of $\mathcal{PE}$, used for communication between elements. As the imported memory units refer to exported memory units of other elements, their names are unchangeable.

The **do...od** section describes the computation involved in $\mathcal{PE}$, modeled by a non-deterministic choice between code resources with bodies $Code_i$. Hence, $\mathcal{PE}$ is a set of code resources with bodies $Code_i$, operating on local and global memory elements. First, the local and exported memory elements whose values form the *state of* $\mathcal{PE}$ are initialized. Then, repeatedly, enabled code resources from $\{Code_i\}_{i \in I}$ are non-deterministically chosen and executed, typically updating the state of $\mathcal{PE}$. Code resources that do not access each other's memory units and are enabled at the same time can be executed in parallel. This is possible because their sequential execution in any order has the same result and the code resources are taken to be atomic. Atomicity means that, if an enabled code resource is chosen for execution, then it is executed to completion without any interference from the other code resources of the system. The computation terminates if no code resource is enabled, otherwise it continues infinitely.

A useful enabling technique for the code resources of a processing element is denoted $cond \rightarrow \mathcal{A}$, where $cond$ is a boolean condition and $\mathcal{A}$ is given by (7.1).

This means that each code resource body $Code_i = g_i \rightarrow B_i$ has the guard condition strengthened from g_i to $cond \wedge g_i$, hence $\mathcal{A}$ can execute only if $cond$ holds.

The location of a processing element The nature of a NoC implies that computation is performed individually at each tile by the modules that can nevertheless communicate with each other. This implies that, even though a processing element may compute using imported data located at other nodes, the individual memory units and code resources defined within the processing element should be located at that respective node. Hence, if we have the processing element

$$
\begin{aligned}
\mathcal{PE} \quad = \quad |[\quad &\textbf{exp} \quad \{(exp\_mu_1, \Phi_1, \cdots), \cdots, (exp\_mu_n, \Phi_n, \cdots)\}; \\
&\textbf{var} \quad \{(local\_mu_1, \Psi_1, \cdots), \cdots, (local\_mu_m, \Psi_m, \cdots)\}; \\
&\textbf{imp} \quad imp\_mu; \\
&\textbf{do} \ [\!]_{i \in I} \, Code_i \ \textbf{od} \\
&]\!],
\end{aligned}
\tag{7.2}
$$

where $Code_i.loc = \Delta_i, \forall i \in I$, then we have that $\Phi_k = \Psi_j = \Delta_i, \forall i, j, k$ and consequently we define that $\mathcal{PE}.loc = \Phi_k$. Conversely, when $\mathcal{PE}.loc = \alpha$, then all the memory units and processing elements defined within $\mathcal{PE}$ have the same location α.

Modularity We observe that a single memory unit or a single code resource can be represented as a processing element. An exported memory unit (mu, l, Val, val) is defined by the following processing element:

$$
\begin{aligned}
mu\_exp \quad = \quad |[\quad &\textbf{exp} \quad mu : Val; \\
&\textbf{init} \quad mu : = val \\
&]\!]
\end{aligned}
\tag{7.3}
$$

while a local memory unit (mu, l, Val, val) is defined by a processing element as follows:

$$
\begin{aligned}
mu\_local \quad = \quad |[\quad &\textbf{var} \quad mu : Val; \\
&\textbf{init} \quad mu : = val \\
&]\!]
\end{aligned}
\tag{7.4}
$$

In these cases, $mu\_exp.loc = l$ and $mu\_local.loc = l$.

A code resource (c, l, C) is defined by the following processing element:

$$
\begin{aligned}
c\_pe \quad = \quad |[\quad & \\
&\textbf{do} \quad c :: C \ \textbf{od} \\
&]\!]
\end{aligned}
\tag{7.5}
$$

In this case, $c\_pe.loc = l$.

Defining memory units and code resources as processing elements is possible due to the high level of flexibility that the base action system framework displays [12]. We employ this feature in Section 7.4.

7.3.1 Conservative extension

We have established elsewhere [243] that MIDAS is a conservative extension of the action systems [11] formalism, i.e., it can be semantically expressed in terms of action systems. This implies that we can reuse all the mechanisms that action systems have also for MIDAS. In particular, we reuse the modularity technique of parallel composition and the refinement technique of superposition.

Parallel composition Processing elements can run in parallel. This operation is described using the *parallel composition* operator.

Consider two elements $\mathcal{PE}_1$ and $\mathcal{PE}_2$ as given below:

$$
\mathcal{PE}_1 \ = \ |[\quad
\begin{array}{ll}
\textbf{exp} & exp\_mu_1; \\
\textbf{var} & local\_mu_1; \\
\textbf{imp} & imp\_mu_1; \\
\textbf{do} & []_{i_1 \in I_1} Code_{i_1} \\
\textbf{od} &
\end{array}
]| \qquad
\mathcal{PE}_2 \ = \ |[\quad
\begin{array}{ll}
\textbf{exp} & exp\_mu_2; \\
\textbf{var} & local\_mu_2; \\
\textbf{imp} & imp\_mu_2; \\
\textbf{do} & []_{i_2 \in I_2} Code_{i_2} \\
\textbf{od} &
\end{array}
]|
$$

We *assume* that the local memory units of $\mathcal{PE}_1$ and $\mathcal{PE}_2$ have distinct names: $\{local\_mu_{1\,j_1}\}_{j_1 \in J_1} \cap \{local\_mu_{2\,j_2}\}_{j_2 \in J_2} = \emptyset$. If this is not the case, we can always rename a local memory unit to meet this requirement. The exported memory units declared in $\mathcal{PE}_1$ and $\mathcal{PE}_2$ are *required* to have distinct names: $\{exp\_mu_{1\,l_1}\}_{l_1 \in L_1} \cap \{exp\_mu_{2\,l_2}\}_{l_2 \in L_2} = \emptyset$. The *parallel composition* $\mathcal{PE}_1 \,||\, \mathcal{PE}_2$ of $\mathcal{PE}_1$ and $\mathcal{PE}_2$ has the following form:

$$
\mathcal{PE}_1 \,||\, \mathcal{PE}_2 \ = \ |[\quad
\begin{array}{ll}
\textbf{exp} & exp\_mu; \\
\textbf{var} & local\_mu; \\
\textbf{imp} & imp\_mu; \\
\multicolumn{2}{l}{\textbf{do}\ Code_1 \,[]\, Code_2 \ \textbf{od}}
\end{array}
]| \tag{7.6}
$$

where $exp\_mu = exp\_mu_1 \cup exp\_mu_2$, $local\_mu = local\_mu_1 \cup local\_mu_2$, $imp\_mu = (imp\_mu_1 \cup imp\_mu_2) \setminus exp\_mu$, $Code_1 = []_{i_1 \in I_1} Code_{i_1}$, and $Code_2 = []_{i_2 \in I_2} Code_{i_2}$. The initial values and locations of the memory units, as well as the code resources in $\mathcal{PE}_1 \,||\, \mathcal{PE}_2$ consist of the initial values, locations, and the code resources of the original elements, respectively. The well-definedness of $\mathcal{PE}_1 \,||\, \mathcal{PE}_2$ is ensured by the fact that all its memory units have unique names. Thus, the exported and local memory units of $\mathcal{PE}_1$ and of $\mathcal{PE}_2$ have distinct names and, moreover, the local memory units of $\mathcal{PE}_1$ can always be renamed in order not to be homonym with the exported memory units of $\mathcal{PE}_2$ (and vice versa). The binary parallel composition operator '$||$' is associative and commutative and thus extends naturally to the parallel composition of a finite set of elements.

Parallel composition and locations If a node hosts several processing elements, then they can be seen as individual elements or as a bigger processing element consisting of the parallel composition of all individual ones. In both cases the global memory units are assumed to be monitored by the router at that node. These memory units consist of the exported (*exp_mu*) and imported (*imp_mu*) memory units of all the processing elements similar to those in the parallel composition (7.6): $exp\_mu \cup imp\_mu$. The parallel composition is independent of the location of its component elements. For a node, all its processing elements are running in parallel with each other having the same location. For the entire mesh, all the processing elements at all the nodes are running in parallel even though their locations are distinct.

Refinement Refinement is a technique for system development. It ensures that a high-level system specification can be transformed by a sequence of correctness preserving steps into an executable and more efficient system that satisfies the original specification. When discussing refinement techniques for a system $\mathcal{PE}$ we refer to the *state* and the *behavior* of $\mathcal{PE}$, where the state of $\mathcal{PE}$ is given by the memory unit values and the behavior of $\mathcal{PE}$ is defined as the set of state sequences that correspond to all the possible executions of $\mathcal{PE}$. In this context, we say that the system $\mathcal{PE}_1$ is *superposition* refined [13] by the system $\mathcal{PE}_2$, denoted by $\mathcal{PE}_1 \sqsubseteq \mathcal{PE}_2$, when the behavior of $\mathcal{PE}_1$ is still modeled by $\mathcal{PE}_2$ and the new behavior introduced by $\mathcal{PE}_2$ does not influence or take over the behavior of $\mathcal{PE}_1$. This means that new memory units and code resources can be added in $\mathcal{PE}_2$, in addition to those of $\mathcal{PE}_1$ and code resources of $\mathcal{PE}_2$ can be modifications of the ones in $\mathcal{PE}_1$, but in such a manner that they do not modify or take over the state evolution of $\mathcal{PE}_1$. For proving the superposition refinement [13] between $\mathcal{PE}_1$ and $\mathcal{PE}_2$, we need to express an abstraction relation between the memory units of $\mathcal{PE}_1$ and $\mathcal{PE}_2$ and to discharge a set of logical properties.

7.3.2   Enabledness

Location-aware computing expands the modeling space of applications as well as restricts the possible executions of these. We model the latter by strengthening the guard conditions used to select code resources for execution hence, code resources will execute less often. The strengthening is modeled by a new boolean condition named *location guard*, conjuncted to the guard condition. The location guard is described based on three predicate types, as detailed below.

Access predicate Location-awareness in our context implies that the imported memory units and the code resource that imports them can have identical or distinct locations. Assume that the names of all the imported memory units used by a code resource $(c, \{\rho\}, C)$ are in the set imp_mu_C. To model the *tile accessibility* of the code resource c, we define a function $cell : Code \times Tiles \rightarrow \mathcal{P}(Tiles)$ depending on the code resource and its location. The cell comprises the set of accessible locations for each code resource c at a certain location $\{\rho\} \subseteq Tiles$.

To model that the memory units imp_mu_C are accessible to the code resource $(c, \{\rho\}, C), \rho \in Tiles$, we define the *access predicate* denoted $access(c@\rho)$ as follows:

$$access(c@\rho) \quad \widehat{=} \quad \forall mu \in imp\_mu\_C \cdot (\exists \alpha \in cell(c, \rho) \cdot (mu \in \alpha.memoryunit) \wedge \tag{7.7}$$
$$(mu.iloc \neq \emptyset \Rightarrow \alpha \in mu.iloc))$$

The access predicate verifies that, for each imported memory unit accessed by the code resource c, there is a location α in the cell of c that contains a memory unit with this name. We can also interpret this rule as follows: For any memory unit named $mu, mu \in imp\_mu\_C$, we choose the location of one of its replicas so that this location is in the cell of the code resource. Furthermore, if the imported memory unit *mu* is specified together with its desired locations of import ($mu.iloc \neq \emptyset$), then the location α is one of the desired locations of import *mu.iloc*.

We note that the cell of the code resource is defined as a function of the application running on the mesh. The accessible locations or tiles for $(c, Tiles_{m \times n}(i, j), C)$ can be formed of the direct neighbors to whom communication is fastest: $cell(c, Tiles_{m \times n}(i, j)) = \{Tiles_{m \times n}(i - 1, j), Tiles_{m \times n}(i + 1, j), Tiles_{m \times n}(i, j - 1), Tiles_{m \times n}(i, j + 1)\}$. However, this is not postulated because, depending on the application the most accessible tiles can be other than the direct neighbors.

Readiness predicate Each location or tile of a NoC mesh supporting a running application can be in either of two states: active or failed. Therefore we partition *Tiles* into the set of active tiles $Tiles_{active}$ and the set of failed tiles $Tiles_{failed}$: $Tiles = Tiles_{active} \cup Tiles_{failed}$, where $Tiles_{active} \cap Tiles_{failed} = \emptyset$.

For NoCs there is no recovery from failed tiles. In this sense, NoC tiles resemble wireless sensors that are cheap and many and can be deployed into the environment to monitor. The failure of a single such sensor does not affect the wireless network of sensors since another sensor can take over and other sensors can be deployed if the network efficiency decreases significantly. However, NoC tiles cannot be replaced once they have failed, hence the resources placed there are simply lost. Therefore, we will allow a code resource $(c, \{\rho\}, C)$ to execute only when this resource and all its imported memory units are located at active nodes. The locations of the code resource and of all its imported memory units are denoted *locations*(c). The *readiness predicate ready*(c) is then defined as $ready(c) = locations(c) \subseteq Tiles_{active}$.

Location guard The *guard* of the code resource $(c, \{\rho\}, C)$ is defined as

$$gd(c, \{\rho\}, A) \mathrel{\widehat{=}} lg(c@\rho) \wedge g(C), \tag{7.8}$$

where

$$lg(c@\rho) \mathrel{\widehat{=}} access(c@\rho) \wedge ready(c), \tag{7.9}$$

and $g(C)$ is the guard condition. A code resource $(c, \{\rho\}, C)$ of a processing element is therefore said to be *enabled*, if its guard $gd(c, \{\rho\}, C)$ evaluates to *true*. A code resource can be chosen for execution only if it is enabled.

Integrity condition A well-defined condition is needed for modeling replication in a framework that requires that all the memory units and code resources have unique names within a processing element. We achieve this by requesting that only one memory unit named *mu* and only one code resource named *c* exist at a certain location $\alpha \in Tiles$. We define the functions $no\_mu : Memory \times Tiles \rightarrow PosInt$ and $no\_coderes : Code \times Tiles \rightarrow PosInt$ where *PosInt* is the set of positive integers including 0. The function no_mu records for every memory unit name *mu* in *Memory* and every tile α in *Tiles* the number of memory units having the name *mu* and located at α. The function $no\_coderes$ records for every code resource name *c* in *Code* and every tile α in *Tiles* the number of code resources having the name *c* and located at α. The integrity conditions are then expressed as:

$$\forall mu \in Memory, \forall \alpha \in Tiles \cdot no\_mu(mu, \alpha) \leq 1 \qquad (7.10)$$

$$\forall c \in Code, \forall \alpha \in Tiles \cdot no\_coderes(c, \alpha) \leq 1 \qquad (7.11)$$

Thus, $no\_mu(mu, \alpha) = 1$ means that a memory unit mu is located or has a replica at α and $no\_mu(mu, \alpha) = 0$ means that there is no memory unit mu located or with a replica at α. We employ the usage of the integrity conditions within the location guard below.

7.3.3   Location-updating code

Replication A classical means to compensate for tile failure is that of replicating resources at various tiles. A *replicated resource* is by definition a resource whose location has more than one element. More precisely, the replicas of a memory unit mu located at $\Gamma = \{\alpha_1, \alpha_2, ..., \alpha_n\}$, $\Gamma \subseteq Tiles$ have the same name, type, and value, but different locations. The replicas of a code resource (c, loc, C) located at Γ have the same name and body, but different locations.

There are two ways to create replicas for memory units and code resources. We can either declare the resources as replicated or we can update their location via code resources during the execution of the processing element. In the latter case, consider that we have a memory unit mu and a code resource (c, loc, C). We create other replicas of this memory unit at the location Γ_1, and of this code resource at location Γ_2, $\Gamma_1, \Gamma_2 \subseteq Tiles$ using special replication code resources denoted $copy(mu, \Gamma_1)$ and $copy(c, \Gamma_2)$, respectively. The details of these definitions are shown in [243]. Essentially, the *copy* resources append the locations Γ_1, Γ_2 to the existing locations of mu and c, respectively: $mu.loc := mu.loc \cup \Gamma_1$ and $c.loc := c.loc \cup \Gamma_2$. A similar code resource is defined for replicating the processing element, in the form $copy(\Gamma)$, see [243].

Mobility Another useful code resource for dealing with efficiency issues and tile failure is one that moves memory units, code resources, and processing elements to various tiles. For the memory unit mu and the code resource c, $move(mu, \alpha_1, \alpha_2)$ refers to the change of location of (a copy of) mu from α_1 to α_2 and similarly $move(c, \beta_1, \beta_2)$ refers to the change of location of (a copy of) c from β_1 to β_2. For moving a (copy of a) processing element from δ_1 to δ_2 we use the form $move(\delta_1, \delta_2)$. The exact definitions are shown in [243].

Location guard for *copy* **and** *move* The location guards of *copy* and *move* are slightly different than those for ordinary code resources, displayed in (7.9). While the detailed differences are documented in [243], we put forward here the extra *integrity predicate* needed to be conjuncted in the location guard with the access predicate and the readiness predicate. Since *copy* and *move* add resources to a destination tile α, we need the integrity predicate to ensure the uniqueness of a resource at a certain location, as described in (7.10) and (7.11). Thus, the integrity predicate referring to copying or moving the memory unit mu to α is in the form $integrity(copy\_or\_move(mu,...))\widehat{=}no\_mu(mu,\alpha) = 0$. Similarly, the integrity predicate referring to copying or moving the code resource c to α is in the form $integrity(copy\_or\_move(c,...))\widehat{=}no\_coderes(c,\alpha) = 0$. For the processing element, a conjunction of the integrity predicates for copying or moving all the memory units and code resources defined within the element form the integrity predicate referring to copying or moving it. The location guard is then defined as $lg(copy(mu,\{\alpha\})@\rho) = access(copy(mu,\{\alpha\})@\rho) \wedge ready(copy(mu,\{\alpha\})) \wedge integrity(copy(mu,\{\alpha\}))$ and similarly for the other resources.

7.3.4 Updating replicated memory units

An extra conjunction to the location guard refers to a necessary condition for updating replicated memory units. Assume a code resource $(c,\{\rho\},C)$ that updates the replicated memory units $\{mu_1,...,mu_p\}$. In this case we need to model that all the replicas of $\{mu_1,...mu_p\}$ are updated simultaneously to the same value. Based on the atomicity assumption and our MIDAS semantics [243], the update is performed to all the copies in the cell $cell(c,\rho)$, hence we need to enforce an extra predicate $updatereplicated(mu_1,...,mu_p) = \forall i \in \{1,...,p\} \cdot mu_i.loc \subseteq cell(c,\rho)$. For code resources c updating the memory units $\{mu_1,...,mu_p\}$, we then add as extra-conjunction the predicate $updatereplicated(mu_1,...,mu_p)$.

7.4 Placing and replacing resources

In this section we describe our proposed placing algorithm. In addition, we describe alternative methods that a once-placed application can employ toward self-replacing when the need occurs.

7.4.1 The placing algorithm

A module $\mathcal{M}$ (a data unit, code resource, or processing element) is called *unplaced* if its location is empty: $\mathcal{M}.loc = \emptyset$. Assume we have an arbitrary application $\mathcal{A}$, specified using MIDAS, to place on the tile mesh *Tiles*. We further assume that $\mathcal{A}$ consists of n unplaced processing elements $\mathcal{PE}_i, 1 \leq i \leq n$ and of m unplaced memory units $\mathcal{MU}_j, 1 \leq j \leq m$, running in parallel: $\mathcal{A} = \mathcal{PE}_1 \parallel ... \parallel \mathcal{PE}_n \parallel \mathcal{MU}_1 \parallel ... \parallel \mathcal{MU}_m$ so that $\mathcal{PE}_i loc = \emptyset, \mathcal{MU}_j.loc = \emptyset, 1 \leq i \leq n, 1 \leq j \leq m$. The required throughput among any two communicating modules together with the initial tile for placing, $Tiles(i_0, j_0)$ (or $Tiles(i_0, j_0, k_0)$ in case of a three-dimensional mesh) are also given. We define the notion of *available tile* as a tile which still accepts modules to be placed there. This is an abstract-enough definition to cover various cases, such as placing only one module at a tile, or placing a module to a tile only if the size of that particular module is suitable for the tile. Here we assume that we place one module at one tile and that the initial tile is available. We define the boolean condition returning whether a location α, $\alpha \in Tiles$ is available as $avail(\alpha) \hat{=} (\alpha.memoryunit = \emptyset \wedge \alpha.coderesource = \emptyset)$.

Communication cost For an efficient placing we first define the distance between any two tiles in a two- or three-dimensional mesh as the number of tile hops among the two. Thus, $d(Tiles(i,j), Tiles(k,l)) \hat{=} |i-k| + |j-l|$. Similarly, $d(Tiles(i,j,k), Tiles(l,r,q)) \hat{=} |i-l| + |j-r| + |k-q|$. Depending on the application, the three-dimensional distance can also be weighed on the vertical dimension, e.g., $d(Tiles(i,j,k), Tiles(l,r,q)) \hat{=} |i-l| + |j-r| + 2|k-q|$. Here, we use the previous definition of the three-dimensional distance.

Given the throughput between any two modules (the non-communicating ones have throughput 0), we define the communication cost between any two modules $\mathcal{M}_1$ and $\mathcal{M}_2$ as the product between their relative distance and their required throughput: $CommCost(\mathcal{M}_1, \mathcal{M}_2) \hat{=} thrput(\mathcal{M}_1, \mathcal{M}_2) \times d(\mathcal{M}_1.loc, \mathcal{M}_2.loc)$. Based on this definition, the communication cost of an application having p modules, $\mathcal{A} = \mathcal{M}_1 \| ... \| \mathcal{M}_p$ is defined as $CommCost(\mathcal{A}) \hat{=} \sum_{i=1,...,p} (\sum_{j=1,...,p} (CommCost(\mathcal{M}_i, \mathcal{M}_j)))$. This definition follows the ones used in other approaches in the literature (see for example [278]) and it has been one of the main metrics for application mapping evaluation.

Algorithm In order to place the modules of $\mathcal{A}$ we propose the following algorithm.

1. We collect the list of communicating modules based on the MIDAS model, together with the required throughput for every two modules.

2. Based on the list from step 1, we determine a *communication graph* that illustrates the communication patterns between all the modules of the application, together with the required throughput on every edge (typically in Mbit/s).

3. Based on this graph we identify the module $\mathcal{M}$ with the biggest number of communication peers and place it at the initial location $Tiles(i_0, j_0)$ (or $Tiles(i_0, j_0, k_0)$): $\mathcal{M}.loc := Tiles(i_0, j_0)$ (or $\mathcal{M}.loc: = Tiles(i_0, j_0, k_0)$).

4. Based on the graph we also identify at most four (or six for the three-dimensional case) communication peers $\mathcal{M}_1, ..., \mathcal{M}_h$, $h = 4$ ($h = 6$), of $\mathcal{M}$ that have the biggest throughput requirements with $\mathcal{M}$ and place them at the direct neighbor locations with respect to $\mathcal{M}$:

$$
\begin{aligned}
&avail(Tiles(i_0, j_0 - 1)) \rightarrow \mathcal{M}_1.loc: = Tiles(i_0, j_0 - 1), \\
&avail(Tiles((i_0, j_0 + 1)) \rightarrow \mathcal{M}_2.loc: = Tiles(i_0, j_0 + 1), \\
&avail(Tiles((i_0 - 1, j_0)) \rightarrow \mathcal{M}_3.loc: = Tiles(i_0 - 1, j_0), \\
&avail(Tiles((i_0 + 1, j_0)) \rightarrow \mathcal{M}_4.loc: = Tiles(i_0 + 1, j_0)
\end{aligned}
\tag{7.12}
$$

We place similarly for the three-dimensional case. We also differentiate between placing memory units $\mathcal{MU}_j$ and processing elements $\mathcal{PE}_i$ for the three-dimensional case.

5. For each unplaced module we find an available location to place it that ensures the minimum communication cost of the module with its communication peers. The placing order of the modules is given by the decreasing number of communication peers deduced from the communication graph.

The generated placement of the modules results in the lowest possible total communication cost of the placed application. At step 4 we observe that the four (six) modules are placed only if the direct neighbor tiles are available. If not, then these communication peers of M will be placed according to the next best strategy given by step 5. However, the initial location should be so chosen that the communication peers of M have available tiles close to M. At step 4 we also need to take into account the 3D NoC structure, such as what type of layers does it consist of. It can be the case that the NoC is seen as a homogenous 3D cube where it does not matter where memory units and processing elements are placed or it could be that each layer of the NoC is meant to contain only one type of module (either memory units or processing elements) and hence, we need to place them according to their type.

We consider the algorithm above as the skeleton of a special processing element, that we name *Placer*. The exact form of this processing element is outside the scope of this chapter, but we discuss it further in the next section.

Applying the algorithm Assume we have an application A with the processing elements $FetchUnit(FU)$, $DecodeUnit(DU)$, $ProgramCounter(PC)$, $HazardUnit(HU)$, $ExecutionUnit(EU)$, and $WriteBackUnit(WBU)$ running in parallel with the memory units $InstructionMemoryUnit(IMU)$, $RegisterAccessUnit(RAU)$, and $MemoryAccessUnit$ (MAU): $A = FU \parallel DU \parallel PC \parallel HU \parallel WBU \parallel IMU \parallel RAU \parallel MAU$.

Further assume we have determined the following list of communicating modules together with the required throughput for each pair:

$$
\begin{aligned}
HU &\Rightarrow DU, thrput(HU,DU) = 200 \\
DU &\Rightarrow PC, thrput(DU,PC) = 180 \\
DU &\Rightarrow RAU, thrput(DU,RAU) = 200 \\
DU &\Rightarrow FU, thrput(DU,FU) = 400 \\
DU &\Rightarrow EU, thrput(DU,EU) = 400 \\
EU &\Rightarrow MAU, thrput(EU,MAU) = 400 \\
MAU &\Rightarrow WBU, thrput(MAU,WBU) = 150 \\
IMU &\Rightarrow FU, thrput(IMU,FU) = 400 \\
WBU &\Rightarrow RAU, thrput(WBU,RAU) = 150 \\
PC &\Rightarrow EU, thrput(PC,EU) = 180
\end{aligned}
\tag{7.13}
$$

Based on this list we determine the graph in Figure 7.3. The graph shows that *DU* is the element communicating with the biggest number of elements (5) and hence should be the module first placed, at the initial location. After that, *EU* communicates with three other elements, the remaining modules except *IMU* have two other elements to communicate with, and *IMU* only communicates with one other element.

If we are working with a two-dimensional mesh $Tiles_{4\times7}$, then a reasonable placement of the modules is illustrated in Figure 7.4. As *DU* has five modules to communicate with and only four tiles as neighbors, it cannot have all five elements

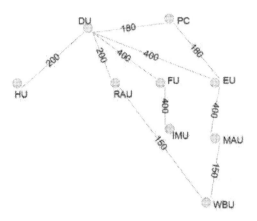

FIGURE 7.3
A communication graph.

around it. As *PC* has the smallest throughput with *DU*, it is the one not placed as a direct neighbor but instead *HU*, *FU*, *EU*, and *RAU* become the direct neighbors of *DU* on $Tiles_{4 \times 7}$. If the initial location was given as $Tiles(3,2)$, then we have $DU.loc := Tiles(3,2)$, $HU.loc := Tiles(3,1)$, $FU.loc := Tiles(2,2)$, $EU.loc := Tiles(4,2)$, and $RAU.loc := Tiles(3,3)$. We then choose $PC.loc := Tiles(4,1)$ since *PC* communicates with *EU* as well. The remaining three placements are decided based on the minimal communication costs among the modules: $MAU.loc := Tiles(4,3)$, $WBU.loc := Tiles(3,4)$, and $IMU.loc := Tiles(1,2)$.

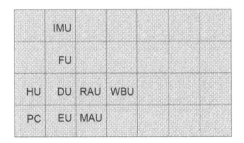

FIGURE 7.4
2D placement example.

Now assume that we have a three-dimensional mesh $Tiles_{3 \times 4 \times 2}$, where we can place the memory units on a distinct layer (the upper one) than the processing elements. In this case a reasonable placement of the elements is illustrated in Figure 7.5.

The memory units *IMU*, *RAU*, and *MAU* should now be placed in the memory unit layer. If we apply the algorithm with the initial location $Tiles(2,2,2)$, ($DU.loc := Tiles(2,2,2)$), then we place *RAU* right above *DU*, $RAU.loc := Tiles(2,2,1)$, and the remaining communication peers of *DU* at the direct neighbor loca-

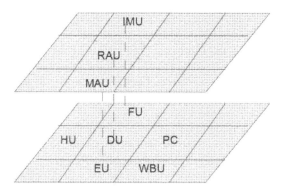

FIGURE 7.5
3D placement example.

tions: $HU.loc: = Tiles(2,1,2)$, $FU.loc: = Tiles(1,2,2)$, $EU.loc: = Tiles(3,2,2)$, and $PC.loc: = Tiles(2,3,2)$. We observe that the communication cost among DU, PC, and EU is the same (540) if PC is placed at $Tiles(2,3,2)$ or at $Tiles(3,1,2)$, i.e., similarly as in the two-dimensional case. The placement of IMU right above FU is obvious, since the former communicates only with the latter: $IMU.loc: = Tiles(1,2,1)$. MAU communicates with both EU and WBU, but the throughout is bigger with EU, hence we place it right above it, $MAU.loc: = Tiles(3,2,1)$, while WBU is placed next to EU, for best communication cost with MAU: $WBU.loc: = Tiles(3,3,2)$. WBU also communicates with RAU, but at the same throughput (150) as with MAU, hence placing WBU at $Tiles(2,3,2)$ instead of PC does not improve on the total communication cost.

It is important to observe that our placing strategy is based on the MIDAS model, that is, on an abstract and non-deterministic type of specification. However, when placing the modules on the NoC, they would have been refined to an implementable and deterministic specification which can be tested for the amount of traffic one communication link actually supports at runtime. This means that detailed decisions of where to place a memory unit such as RAU are based on more criteria than just the minimum communication cost among peers. We investigate several ideas related to runtime-based placing in the following section.

7.4.2 Replacing

In the following we describe three techniques we have identified with respect to modifying the placement of an application on a NoC, here referred to as *replacing*. The first approach discusses reapplying the *Placer* algorithm when the need occurs, the second approach concerns the replication of resources, both memory units and processing elements, and the third approach concerns tile failures on a NoC.

Centralized replacing In case there are some changes to the application requirements (e.g., the throughput), the placing algorithm presented in Section 7.4.1 has

to be rerun. This is also the case when some modules have changed their location, as we describe shortly or when the NoC mesh is reformatted. To model
such occasional rerunning of the placing algorithm, we need to first model the enabling (boolean) condition *place_again* and to enable the *Placer* processing element
only when *place_again* holds: *place_again* → *Placer*. Then, we compose in parallel this conditional reexecution of the placing algorithm with the application $\mathcal{A}$:
$\mathcal{A} \,||\, (place\_again \to Placer)$.

Placer consists of the algorithm described in the previous section, that initializes
the locations of all the modules to $\emptyset$ and runs until all the module locations are distinct
from $\emptyset$. One extra code resource to add in *Placer* is that of disabling *Placer*:

$$\bigwedge_{i=1,\dots,p} \mathcal{M}_i.loc \neq \emptyset \to place\_again := \neg place\_again$$

As the computation in $\mathcal{A}$ does not take the location (tile) into account, it is clear
that *Placer* adds new code resources to $\mathcal{A}$ that do not modify the state of $\mathcal{A}$ and
do not take over the computation in $\mathcal{A}$ (*Placer* obviously terminates). Hence, $\mathcal{A} \sqsubseteq$
$(\mathcal{A} \,||\, (place\_again \to Placer))$.

This refinement means that we have separated the computation (modeled by $\mathcal{A}$)
from the replacing (modeled by *place_again* → *Placer*) in such a manner that the replacing is correctly applied to the application. We do not show the formal proof here,
since we did not elaborate the details of *Placer* in this chapter. Details on refinement
proofs are shown elsewhere [147, 243].

We also note that *place_again* → *Placer* proposes a centralized replacing solution since it is based on a global data structure (the communication graph) as well
as on the decreasing order of the number of communication peers for the modules to
place. In the following we also discuss distributed replacing solutions.

Distributed replacing The placement strategy described in Section 7.4.1 is based
on the communication peer relationships, favoring an advantageous placing to the
modules with many communication peers. These modules are placed with the communication peers as close as possible on the NoC to ensure efficiency. However, this
might not be the optimal solution as these modules are candidates for bottlenecks
due to possibly many communication requests simultaneously. As this situation can
occur only if the communication peers have a high volume of communication and request it performed simultaneously, measuring the communication traffic is necessary
at runtime.

For modeling this, we define as threshold a certain limit *commlimit* of the communication traffic for each module $\mathcal{M}$. When this threshold is overcome, *replicas*
of the highly communicating modules can be made. We define a local memory unit
commtraffic within each module of an application, memory unit or processing element, because each can become overloaded by communication. Then, each code
resource of each module strengthens its guard with the condition *commtraffic* <
commlimit, where *commlimit* is modeled as a constant memory unit defined for each
module placed on a tile. In addition, we define a new code resource:

$$commtraffic \geq commlimit \to \{\alpha : \in Tiles | avail(\alpha)\}; copy(\alpha) \tag{7.14}$$

Here ';' models the sequential composition of the code resources $\{\alpha : \epsilon$ $Tiles|avail(\alpha)\}$ and $copy(\alpha)$. The former code resource models the non-deterministic assignment of an arbitrary location from *Tiles* to α so that $avail(\alpha)$ holds. The new code resource (7.14) thus models that the module is replicated at an available location when the communication threshold is reached or overcome. Many other restrictions can be imposed with respect to how α is chosen. One approach is to have α as close to $M.loc$ as possible, while another (non-exclusive) one is to choose α from the same layer. The guard condition of the new code resource ensures that the latter is only enabled when there is an available location α in *Tiles*. Otherwise the code resource cannot be chosen for execution. We also note that, according to the location guard of the code resources (discussed in Section 7.3), α needs to belong to the cell of the *copy* code resource and be an active tile where there are no other code resources or memory units bearing the same name as those defined within M. We illustrate the skeleton of the modified module $M_{runtime}$ in (7.15).

$$
M_{runtime} \;\;\widehat{=}\;\; [\![\quad \cdots
$$

$$
\begin{aligned}
&\textbf{var} \quad \{\cdots, (commtraffic, \cdots\} ; \cdots \\
&\textbf{do}\; commtraffic < commlimit \rightarrow \cdots \\
&[\!]\cdots \\
&[\!]commtraffic \geq commlimit \rightarrow \\
&\qquad \{\alpha : \epsilon\, Tiles|avail(\alpha)\}; copy(\alpha) \\
&\textbf{od}
\end{aligned}
$$

$$
]\!]
$$

(7.15)

We observe that the replicated memory units have to be kept consistent with the original and the communication peers should be able to communicate with either of the original and replicated module. The first issue is addressed by the $updatereplicated(mu_1, ..., mu_p)$ conjunction from the location guard of any code resource replicating memory units $mu_1, ..., mu_p$ (7.3.4). If not all the replicas of a memory unit can be accessed by the updating code resource, then the latter is not enabled and the update does not take place. Hence, all the replicas are consistent in our models. The second aspect is embedded in the our framework where a code resource is choosing replicated and imported memory units in its cell, by name, to work with. If we define the cells conveniently enough, e.g. the physically closest tiles, then we have solved the problem and the application has become more efficient.

The modified module $M_{runtime}$ of the application A at runtime has stronger guards than the original module M as well as an extra code resource (7.14) that does not modify or take over the computation in A. Hence, $M \sqsubseteq M_{runtime}$, meaning that our replacing is correctly applied to A.

We also observe that the modified modules $M_{i_{runtime}}, i = 1, ..., p$ provide a distributed replacing where each module takes care of the local bottlenecks based on runtime measurements. The memory units *commlimit* and *comtraffic* are local to each module and thus need to be renamed in the parallel composition $A = M_{1_{runtime}} \| ... \| M_{p_{runtime}}$ to $commlimit_i$ and $comtraffic_i$ respectively, where $i = 1, ..., p$.

Tile failure With respect to tile failures, NoC tiles cannot be replaced once they have failed, hence the resources placed there are simply lost.

There are at least two directions for improving this situation. First, if we can define some monitoring code that predicts the tile failure, then we could move the resources placed there to another (available) location and the NoC would continue to work without any loss of efficiency. Second, if we can keep a list of all the resources and their placement on tiles in a safer tile, perhaps one able to tolerate various faults, then upon detection of a failed tile, we can simply replicate the resources placed there to a tile that is available. We discuss these two directions below.

The first approach is very similar to the distributed replacing above, as we add a code resource in each module, of the form

$$tile\_may\_fail(\mathcal{M}.loc) \rightarrow \{\alpha: \in Tiles | avail(\alpha)\}; move(\alpha) \tag{7.16}$$

where $tile\_may\_fail(\mathcal{M}.loc)$ returns a boolean value depending on various sensor measurements at runtime for the module $\mathcal{M}$ and the new location α to move the processing element to is chosen similarly as in the distributed replacing. We have that $\mathcal{M} \sqsubseteq \mathcal{M}_{failmove}$, where $\mathcal{M}_{failmove}$ is the module $\mathcal{M}$ with the code resource (7.16) added. The refinement holds because we only have the extra code resource in $\mathcal{M}_{failmove}$ that does not modify or take over the computation in $\mathcal{A}$. This means that our replacing is correctly applied to $\mathcal{A}$. We also observe that the modified modules $\mathcal{M}_{i_{failmove}}, i = 1, ..., p$ provide a distributed replacing where each module takes care of the local potential tile failures, based on runtime sensor measurements.

The second approach provides a centralized runtime replacing based on a safer tile (backup tile), say $Tiles(i, j, k)$, known in advance. We assume that this tile can store all the application modules, which is a different assumption than we used for the placement. Depending on the application and tile mesh, there could be more than one backup tile, but here we just assume the easiest case for simplicity. The modules of the application and their locations are stored at $Tiles(i, j, k)$. The application $\mathcal{A}$ then has an extra processing element *Backup* that is composed in parallel with it: $\mathcal{A} \parallel Backup$. *Backup* addresses the (periodic) checkup for failed tiles:

$$
\begin{aligned}
Backup \quad \widehat{=} \quad \parallel[\quad &\textbf{var} \quad \{(to\_check, Tiles(i,j,k), \mathcal{P}(Tiles), Tiles)\}; \\
&\textbf{do } to\_check \neq \emptyset \rightarrow \\
&\quad (\{\alpha: \in to\_check\}; \\
&\quad (\alpha \in Tiles_{failed} \rightarrow \{\beta: \in Tiles | avail(\beta)\}; \\
&\quad copy(\alpha.memoryunit, \beta); copy(\alpha.coderesource, \beta)); \\
&\quad to\_check: = to\_check \setminus \{\alpha\} \\
&\quad) \\
&\textbf{od} \\
\parallel &
\end{aligned}
\tag{7.17}
$$

We have that $\mathcal{A} \sqsubseteq (\mathcal{A} \parallel Backup)$, because we have the extra code resource in *Backup* that does not modify or take over the computation in $\mathcal{A}$. This means that this replacing type is also correctly applied to $\mathcal{A}$. This replacing is done in a centralized manner as it is based on global knowledge regarding the entire application (stored at $Tiles(i, j, k)$) as well as the entire tile mesh that is tested for failures.

Concluding, all three replacing directions described above assume some monitoring code that runs in parallel with the application. This code either occurs when needed or predicts failures or determines if they have occurred, upon which it replicates or moves resources placed there to a safer (available) tile. This type of monitoring code does not influence or take over the application, acting as a superposition refinement of code. More importantly, this implies that the NoC-running application is enabled to *self-replace*, thus leading to a higher degree of autonomy for Networks-on-Chip.

7.5 Conclusions

In this chapter we propose a middleware language based on formal methods and developed for real networks to be adapted to applications running on NoC. The network nodes become the NoC tiles and many of the issues remain unchanged, such as the need for replication. The focus of the chapter is on how to place modules that communicate often in each other's vicinity for efficiency. This becomes even more interesting if the NoC consists of vertically stacked layers, commonly referred to as 3D NoCs. Furthermore, we analyze three dynamic alternatives for replacing an application, some of which are centralized and some of which are distributed. Based on our base formalism we show that the replacing is correctly applied to the application.

Our future work is considered in two main directions. First, once we have presented our placing algorithm and our replacing proposals, we need to apply them and assess their value on several realistic case studies. One such example is provided by studying H.264, the latest standard of video stream coding [51]. Second, we plan to explore some formal method tools for easing the modeling and analysis tasks as well as preparing our method for a different impact.

One tool to explore is ProB [180], dedicated to verifying models written in the B specification language. The B method [2] is very similar to the original formal method behind MIDAS, i.e., action systems [11]. As the B method has various associated tools, such as ProB and also a theorem prover under continuous development [1], we can develop our models in MIDAS and then translate them into ProB models. The translation has been shown before to be straightforward [315]. Various verification techniques, such as model checking and several animation features are provided in ProB to assist the analysis of a model. We have already explored [299] some ProB facilities for placing pipelined applications, where various other relations are used instead of our communication graph (Figure 7.3). The other tool to explore is the RODIN platform mentioned above. With it, we can specify and prove the correctness of the application modules and also explore the proof that our replacements are correct.

Finally, our conjecture is that replacing can be done by adding monitoring code running in the background of the application without interfering with it. Formally, this is a superposition refinement. However, the most important benefit would be that an application having monitoring code for replacing itself is a step forward toward autonomous NoCs.

8

Self-Adaption in SoCs

H. Zakaria

TIMA CNRS, Grenoble Institute of Technology, UJF, France
Banha High Institute of Technology, Banha, Egypt

E. Yahya

TIMA CNRS, Grenoble Institute of Technology, UJF, France
Banha High Institute of Technology, Banha, Egypt

L. Fesquet

TIMA CNRS, Grenoble Institute of Technology, UJF, France

CONTENTS

In recent years, there has been tremendous growth in silicon integration capacity. It is now possible to have complex systems (e.g., large multiprocessor, memories, IPs, complex clock trees, I/O control units) implemented on a single chip. On the other hand, the design of such complex System-on-Chips (SoCs) is now constrained by many parameters such as speed, energy and robustness to process variability. In fact, controlling the speed and the energy in a complex SoC requires specific power supplies and several clock generators. Furthermore, with the technology shrink, the behavior of such systems that embed several microprocessors and a complex Network-on-Chip (NoC) is no more predictable. In order to reach an acceptable fabrication yield, the clock synchronization based on the assumption that the critical path is shorter than the clock period is impracticable with large SoCs. As a result, the complex SoCs are divided into multiple clock domains. These SoCs are called Globally Asynchronous Locally Synchronous (GALS) circuits. The GALS circuits provide a promising solution for implementing the large and complex SoCs.

As the designers face several challenges in the nanometric era, this chapter targets the following main issues: power consumption, process variability, synchronization (in SoCs) and yield. In order to design a self-adaptable SoC, an analysis and a solution to each of these challenges are presented. Only GALS systems are considered but the proposed solutions are also applicable to simpler designs. For instance, the application of a Dynamic Voltage and Frequency Scaling (DVFS) is applied to make self-adaptable SoC to its constraints in term of energy. In order to be tolerant to the process variability, specific sensors are also proposed to evaluate the fabrication process quality and the local environmental parameters (voltage, temperature) in each clock domain. The outputs of these sensors, smartly combined with the DVFS, provide the required information to determine the appropriate clock frequencies which do not violate the local timing constraint. Therefore the SoC also is able to self-adapt its speed and voltage to the process variations. This implies to feedback the system and to implement a well-suited control in order to efficiently manage the process variability as well as the energy or the speed. This leads to a potential fabrication yield improvement because the system is able to adapt itself to the process variation in order to guarantee the correctness of its behavior. Moreover, the implementation of GALS systems makes mandatory the data synchronization to avoid metastability problems. Therefore different techniques of synchronization are discussed and compared.

8.1 Introduction

Embedded integrated systems in multimedia and telecommunication applications are today low-power systems which increase by two their computational load at each new generation in order to respond to the market evolution. As the systems have reached limits in terms of power consumption, computational efficiency and fabrication yield, the upcoming generations of embedded integrated systems require several drastic technological evolutions. The main problems that we are facing nowadays with the nanometric technologies can be categorized in three main problems (Power Consumption, Process Variability, and Yield).

8.1.1 Power consumption

Rapid development of portable systems like laptops, PDAs, digital wrist watches, and cell phones requires low power consumption and high density ICs, which leads to a surge of innovative developments in low power devices and design techniques. In most cases, the requirements for low power consumption must be met with equally demanding goals for high chip density and high throughput circuits. Hence, the low power digital design and digital ICs have emerged as very active fields of research and development. In this cutting-edge technology era, reduction in the power dissipation is a critical task, especially as the size of transistors is scaled down to increase the transistor density over the silicon chip. Consequently, careful consideration must be given to minimize digital ICs power dissipation without sacrificing its performance.

Power dissipation in CMOS circuits involves both static and dynamic power dissipations. In the submicron technologies, the static power dissipation, caused by leakage currents and subthreshold currents contribute with a small percentage to the total power consumption. This static power consumption occurs when all inputs are held at the same logic level and the circuit is not in changing states. On the other hand, the dynamic power dissipation, resulting from charging and discharging of parasitic capacitive loads of interconnects and devices dominates the overall power consumption. Therefore, traditional low power management techniques for 130nm and 180nm technology nodes were focusing only on reducing the dynamic power.

Advances in CMOS technology allow MOS transistor sizes to be continuously scaled down in progressively smaller technology nodes. As transistor sizes become smaller, supply voltages can be lowered to reduce the power dissipation. In order to achieve high speed with low supply voltages, the threshold voltage must be reduced accordingly. This reduction of the threshold voltages combined to the short channel effect has led to an exponential increase in the transistors leakage current. Thinner gate oxides have also contribute to an increase in gate leakage current as well. Consequently, leakage power is no longer negligible in deep-submicron CMOS technologies. At 90 nm and beyond, leakage power accounts for a significant portion of the total power in high-performance designs as shown in Fig 8.1 [238], therefore its management becomes essential in the ASIC design process. Leakage power is now a growing concern in the overall design process. Unlike dynamic power, which can be managed by reducing switching activity, leakage power exacts its effect as long as the power is on. That's why current process technologies are pushing designers to consider new design methods to reduce both static and dynamic power consumption. In this chapter a survey on different methods that can be used to control these two main types of power consumption will be presented in order to make the chip self-adaptable with respect to its power consumption during the design phase.

8.1.2 Process variability

Variability refers to the unpredictability, inconsistency, unevenness, and changeability associated with a given feature or specification. With the nanometric technologies, variability has become one of the leading causes for chip failures and delayed schedules. In nanometric design flows, variability is associated with design modes, power states, environmental conditions, manufacturing steps, and the behavior of devices and interconnects. Variability affects the entire physical design environment, from power management through timing and signal integrity closure to manufacturability.

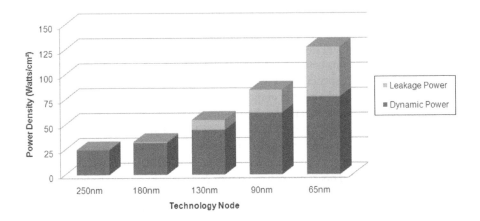

FIGURE 8.1

Static vs. dynamic power with process migration.

"As technology continues to advance deeper into the nanometer regime, the ability to control the process parameters is increasingly difficult. As a consequence, variability has become a dominant factor in the design of ICs."

Davide Pandini
ST Microelectronics

A major problem facing the computer and semiconductor industries is the increasing amount of CMOS process variability. Variability in low-level circuit parameters, such as transistor gate length and gate oxide thickness, complicates system design by introducing uncertainty about how a fabricated system will perform [261]. Although a circuit or chip is designed to run at a nominal clock frequency, the fabricated implementation may vary far from this expected performance.

8.1.3   Yield

Early on, integrated circuit design rules were absolute and finite. The path to yield was fairly simple: comply with all the design rules and yield would follow. Designers did not need to worry too much about what happened in the fabrication after tape-out [228]. In the nanometric era, the game has changed. Yield success is much harder to achieve because of the increased number and complexity of variables affecting manufacturability. The definition of yield itself has changed, now incorporating measures of variable power management, multi-modal performance and circuit integrity. The designer's strategy must shift from simple design rule compliance to the definition and design of the optimal layout for the highest yield.

Suggested Solutions to the Nanometric Challenges

Regarding the previously mentioned motivations, we can conclude the suggested solutions to the different nanometric design challenges in two main points: Firstly, the need for robust power management techniques that control processor speed, power, and energy, secondly, the need to handle the impact of process variability and control the produced amount of process uncertainty. If we succeed to do so, then we could design a SoC at very small geometries which is self-adaptable with respect to all these *nanometric* challenges. Each of these suggested points of solution will be discussed in details in the following sections.

8.2   Power management techniques

The power consumption of modern high-performance systems is becoming a major concern because it leads to increased heat dissipation and decreased reliability. Since all applications running over system processors do not require fast processors to operate at the highest speed all the time, it is possible to slow down processor speed or distribute applications with the processing load on the slowest processors. Moreover, these techniques are able to slow down the fast circuitry and reduce the static power consumption as well as the dynamic power consumption when the maximal performance is not required.

There are three major sources of power dissipation in digital CMOS circuits which are summarized in the following equation:

$$P = P_{dynamic} + P_{shortcircuit} + P_{leakage}$$
$$= \alpha C_L V_{dd}^2 f_{clk} + I_{sc} V_{dd} + I_{leakage} V_{dd} \quad (8.1)$$

The first term represents the switching component, *dynamic power* consumption which occurs when a gate is switching from one logic state to another and is the result of the switching current (needed to charge internal nodes), where C_L is the load capacitance, f_{clk} is the clock frequency and α is the probability that a power consuming transition occurs (the activity factor). The second term, *short circuit power*, consumption is due to the direct-path short circuit current, I_{sc}, which arises when both the NMOS and PMOS transistors are simultaneously active, conducting current directly from supply voltage to ground. Finally, *leakage power* consumption, where $I_{leakage}$ can arise from substrate injection and sub-threshold effects, which is primarily determined by fabrication technology considerations. In the following sections we are going to present in more details how to deal with the dynamic and leakage part of power consumption.

8.2.1 Leakage power management

In today's technology nodes, leakage power is a significant contributor to the total power, as the gate length and threshold voltage are scaled down. Several techniques can be applied at the circuit level to reduce leakage power, including multi-threshold libraries, multiple and dynamic supply voltages, power gating and variable body biasing [165, 314].

Multi-threshold libraries: Usually a certain number of libraries of gates are defined for different implementations of functions, including high-voltage-threshold (HVT), standard-voltage-threshold (SVT) and low-voltage-threshold (LVT) cells, which have different speed and leakage characteristics. Of them, the LVT cells are the fastest and have the highest leakage. They are used by the synthesis and optimization tools in the critical paths. The SVT and HVT gates are used in less-critical paths to reduce leakage power.

Body biasing: Techniques may also be used to control leakage dynamically in sections of the chip that are idle. An adaptive body biasing approach is presented in [298]. The chip area is divided into small blocks, and each block has a replica of the critical path. The delay of the replica is used as an indicator for body-biasing the transistors in that block.

Power gating: Technique wherein circuit blocks that are not in use are temporarily turned off to reduce the overall leakage power of the chip and once they are required for operation again they are activated. These two modes are switched at the appropriate time and in the suitable manner to maximize power performance while minimizing impact to performance. Thus the goal of power gating is to minimize the leakage power by temporarily cutting power off to selective blocks that are not required in that mode. Power gating affects design architecture more compared to the clock gating. For example in some cases when power gating is used, the system needs some form of state retention, such as scanning out data to a RAM, then scanning it back in when the system is reawakened. Thus it increases time delays as power gated modes have to be safely entered and exited. The possible amount of leakage power saving in such low power mode and the energy dissipation to enter and exit such mode introduces some architectural tradeoffs. Shutting down the blocks can be accomplished either by software or hardware.

8.2.2 Dynamic power management

Clock speed and integration density have been used as the only performance metric of electronic devices for a long time. Therefore designers mainly focused their efforts on increasing the transistor density and the clock frequency. The power consumption was of secondary importance. On one hand, with the proliferation of mobile computing and portable devices, reducing power is now one of the major challenges in electronic system design [28, 240]. On the other hand, these systems tend towards high performances in computation and in service integration which increase the power demand and are contradictory to the previous requirement. Although systems such as mobile phones has to support video transmission as well as internet access and some other data intensive applications, reducing power consumption is highly suitable to keep reasonable autonomy and battery weight. Several hardware and software techniques have been developed over the last years to manage the power/energy of consumption. Nevertheless, as devices become much more powerful and sophisticated, power requirements increase continuously. Therefore new power management techniques have been investigated [245].

For current CMOS integrated circuits, the average power consumption P_{ave} and the energy dissipation E is dominated by the switching power that arises from the charging and discharging of the load capacitance and can be expressed as

$$P_{ave} \propto f_{clk} V_{dd}^2 \tag{8.2}$$

$$E \propto C V_{dd}^2 \tag{8.3}$$

where C is the load capacitance, V_{dd} is the supply voltage and f_{clk} is the clock frequency. These equations suggest that minimizing the load capacitance, reducing the supply voltage or slowing the clock can reduce power and energy. While the load capacitance can only be affected during chip design (for example by minimizing on chip routing capacitances and reducing external components access), voltage scalable processor and power controllable peripheral devices make it possible to reduce power and control it by a dedicated digital hardware or by an Operating System (OS). For instance, the OS can control the processor frequency and its voltage (DVFS) and/or put the devices in low-power sleep/idle states (Dynamic Power Management) [189,276]. The power minimization can be achieved by resolving a stochastic optimization problem off-line but this is not always possible. Therefore the optimization has to be performed on-line by the control system dedicated to the power management. These techniques based on DVFS slower the processors speed at a reduced voltage according to the instantaneous computational demand. Moreover, this approach is often combined with the management of the processor and peripheral low-power idle modes. The variety of sensors and actuators and the difficult system identification in such system make difficult the design of the power control system.

As shown in Eq. (8.2) and Eq. (8.3), reducing supply voltage V_{dd} will decreases both the energy and the average power dissipation, while reducing the clock frequency f_{clk} will reduce only the average power dissipation. The reduction in average power is a quadratic function of the voltage V_{dd} and a linear function of the clock frequency f_{clk}. As a result, Dynamic Voltage Scaling (DVS) can be used to efficiently manage the SoC energy consumption [99]. Supply voltage can be reduced whenever slack is available in the critical path. However, this drop will also decrease the computational speed because of the propagation delay of gates, i.e. T_d, which is seriously increasing as V_{dd} approaches the threshold voltage of the device V_t (see Fig 8.2). In this figure an inverter implemented on STMicroelectronics 45nm CMOS technology using three different libraries (i.e. HVT, SVT and LVT) was taken as an example. Fig 8.2 also shows that for the same gate, the LVT implementation provides the smallest propagation delay while the HVT implementation presents the largest amount of delay, as explained in Section 8.2.1. The relation between the propagation delay T_d and the supply voltage V_{dd} can be expressed as:

$$T_d \propto \frac{V_{dd}}{(V_{dd} - V_t)^2} \tag{8.4}$$

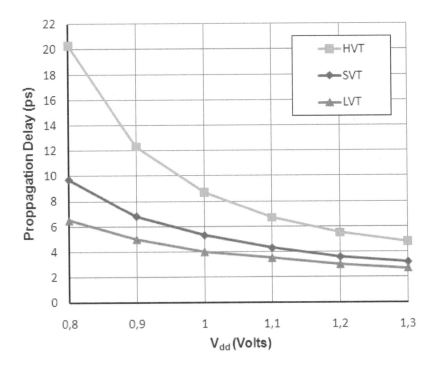

FIGURE 8.2

Propagation delay vs. V_{dd} for an inverter gate in STMicroelectronics 45nm CMOS process using HVT, SVT and LVT libraries.

Controlling the supply voltage is hence a power-delay tradeoff: the power consumption decreases while the delay increases. That is why the supply voltage and the clock frequency have to be controlled together to guarantee the critical path (i.e., the slowest path between two clocked components). If the frequency is too high the data cannot be computed during a clock period, which means that the results sampled by the clocked components will probably be incorrect. Therefore the critical path delay has to be lower than the the clock period [249]:

$$T_{d(critical\ path)} < \frac{1}{f_{clk}} \qquad (8.5)$$

Clearly, it is required to decrease the clock frequency before decreasing the supply voltage and, respectively, increase the supply voltage before increasing the clock frequency. This principle is needed in all systems to guarantee the critical path: either with a hardware solution like some delay lines [77, 78], or with a software technique at least.

Adapting the supply voltage is very interesting when possible but this implies the use of Dynamic Frequency Scaling (DFS) to keep correct the system behaviour. The addition of DFS to DVS is called DVFS and results in simultaneously managing the frequency and voltage which provides a good consumption-performance tradeoff.

As applications do not require the full computational power at any time, it is possible to control speed and energy. In many cases, the only performance requirement is that the task meets a deadline (see Fig 8.3(a)). Such cases create opportunities to run the processor at a lower performance level and achieve the same perceived performance while consuming less energy. Figure 8.3 (b) shows that decreasing the processor clock frequency reduces power consumption but simply spreads the computation out over time, thereby consuming the same total energy as before. Fig 8.3 (c) shows that reducing the voltage level as well as the clock frequency achieves the desired goal of reduced energy consumption at an appropriate performance level [307].

8.2.2.1 DVFS architecture overview

DVFS is a technique that allows the processor to dynamically alter their voltage and speed at runtime under the control of voltage scheduling algorithms. Implementing DVFS in a general-purpose microprocessor system includes three key components:

1. An operating system that can smartly vary the processor speed

2. A regulation loop which generates the minimum voltage required for the desired speed

3. A microprocessor which operates over a wide voltage range

Nowadays, because of the instruction complexity and the access time to memory through different levels of cache, it is extremely hard to predict the execution time of a program and consequently the real speed of a processor. DVFS implementation generally assumes that the speed of the processor is constant (do not care about the cache mechanisms) and takes the worst case. With such an assumption, the processors run faster than the minimal speed enabling the highest energy savings. With a feedback system, it is possible to control with high accuracy the power/energy consumed by the circuit. This is done respecting all the specifications and needs of the applications executed by the processor.

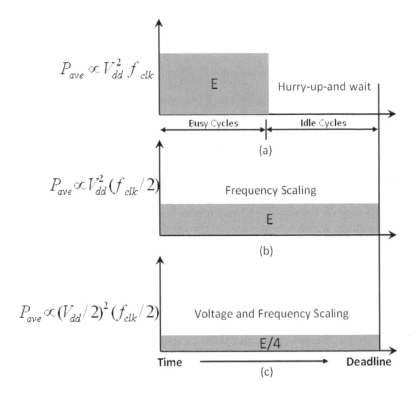

FIGURE 8.3
Energy consumption vs. power consumption.

Figure 8.4 shows the block diagram of the DVFS system. The sensor integrates an instruction counter and the clock is used as a time reference. Because the processor informs the sensor every time an instruction completes, it can calculate the real speed of the processor in Millions Instructions per Second (MIPS), averaged on a period of computation. Indeed, each executed application indicates to the co-processor through a software layer (OS) the speed it requires. The information about the speed can be inserted statically into the code at compile time (API, or Application Programming Interface, function calls in the code) or computed dynamically at run time by the operating system. Therefore, combining the information about the real-time requirements (estimated in MIPS) of the applications and OS running on the microprocessor enables us to create a computational load profile with respect to time. Consequently, using this profile it is possible to apply a fine-grain power/energy management (software) allowing application deadlines to be met. Note that task scheduling is an important issue when using this technique. Sophisticated scheduling policies, such as the one proposed in [336], can easily take advantage of this work by simply interfacing to the hardware features proposed using an API. The co-processor inte-

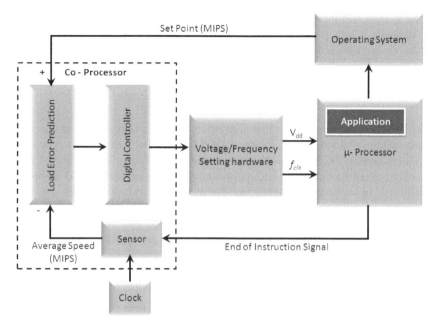

FIGURE 8.4
Voltage/frequency control simplified architecture.

grates with the sensor two main parts, the load error prediction unit and the digital controller part. The load error prediction computes the amount of error in the load profile (i.e. the difference between the set point sent by the OS and the calculated average speed by the sensor). The digital controller part dynamically controls the voltage/frequency scaling hardware which integrates a DC/DC converter to scale the supply voltage as well as the programmable clock generator to control the clock frequency in order to satisfy the application computational needs with an appropriate management strategy. This leads to a lower supply voltage and frequency reducing the system power/energy consumption. Indeed, the lower the processor speed, the lower the power consumption.

There are many ways to control the voltage/frequency scaling hardware (e.g., Proportional Integral Derivative (PID) controller, fuzzy logic compensators). Since the control is performed by the digital controller of the co-processor, this enables us to choose the way the control operates. The control depends on the microprocessor and the DC/DC converter. For each circuit it is necessary to perform a complete characterization of the whole system in order to tune the control parameters at best. For instance, the period of the external sensor clock is crucial since it determines the accuracy of the calculated average speed. Moreover, it determines the system speed response. In fact the co-processor sensor integrates also a register to memorize the number of executed instructions (i.e. the counter output) on a predefined period and determines the average speed on each rising-edge of the clock. If this period is short, the system will be fast but the calculated average speed will not be accurate. On the opposite, a long period leads to a slow system but to a more accurate speed. Therefore, the period used to estimate the speed must be chosen carefully.

8.2.2.2   DC/DC converter

The DC/DC converter is a circuit that converts a voltage source (of direct current) from one voltage to another. Two kinds of DC/DC converter can be used. The first class is a continuous DC/DC converter which provides an accurate supply voltage. Another kind of converter used is a digitally controlled step-converter (Vdd-Hopping converter) that has a better efficiency but discrete output values. A 2-bit and a 4-bit converter are considered in [259] and modeled using VHDL-AMS. A special step is defined to obtain the minimum voltage when the processor is idle. It has been shown that with a continuous DC/DC converter, the speed of the processor exactly follows the set point. This is the most accurate way to control the speed of the processor because the voltage is supposed to be the exact image of the set point. On the contrary, with a voltage hopping converter, the output voltage is discrete and it is not possible to exactly follow the calculated speed. Then, in order to meet the deadlines, the supply voltage is always chosen a little bit higher.

With discrete DC/DC converters it was noticed that increasing the number of bits should lead to improve the efficiency of the regulation and therefore decrease the energy consumption. However, a higher number of bits increases the complexity of the control and decreases the efficiency of the DC/DC converter. There is obviously a tradeoff between the accuracy of the speed set point and the energy cost of the controller and DC/DC converter. The best solution is to jointly define the number of bits as well as the voltage values.

Figure 8.5 presents the energy consumed by a system for different architectures and for a given application profile. It is clear that the DVFS technique allows reducing the energy consumed by the micro-processor. As far as the speed regulation is concerned, the continuous DC/DC converter gives the best regulation of the processor speed and then the lowest consumption of the processor alone. However, the DC/DC converter efficiency must be considered to calculate the consumption of the whole system. With the continuous DC/DC converter, the whole system consumption is higher than with the step converters. As a matter of fact, the total energy consumption using the 4-bit regulator is lower than a system without DVS by 33% [259]. Therefore, the choice of the number of bits is important and depends on the targeted system. In fact, increasing the number of bits leads to improve the efficiency of the regulation and therefore decrease the energy consumption. However, a higher number of bits increases the complexity of the control and decreases the efficiency of the DC/DC converter. Therefore, there is obviously a tradeoff between the tracking accuracy of the speed set point and the energy cost of the controller and DC/DC converter. The best solution is to jointly define the number of bits as well as the voltage values.

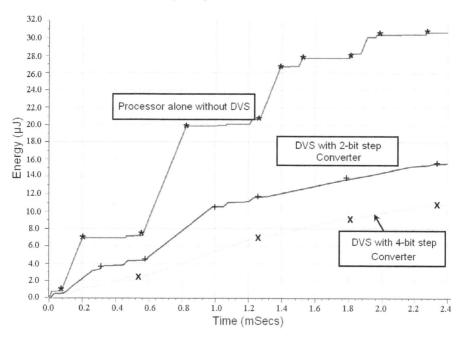

FIGURE 8.5

Energy consumption of the processor alone for different discrete DC/DC converter architectures for a given application [259].

8.2.2.3 Clock generator

The application of DFS to a system requires the use of a source for generating adjustable clocks. For example these clocks can be derived from voltage controlled oscillators (VCOs), which are a part of a phase locked loop (PLL). However, VCOs have a limited operating range and required a stabilization time when changing the frequency [47]. Another solution is to use a standard clock divider, but this will make the time resolution coarser, due to counting integer periods of the input frequency [287]. In addition, they give regular time step which implies irregular frequency step (usually frequency step follows 1/x curve). Self-timed rings are considered as a promising solution for generating clocks [326]. In [92] they are efficiently used to generate high-resolution timing signals. Their robustness against process variability in comparison to inverter rings is proven in [120].

Self-Timed Ring Figure 8.6.a shows the structure of the ring stage. It is composed of a C-element (Muller gate) and an inverter. Table 8.1 shows the truth table of the C-element, where it sets its output y to the input values (a and b) if the inputs are equal and holds its output otherwise. In each ring stage, the input which is connected to the previous stage is marked F (Forward) and the input which is connected to the following stage is marked R (Reverse), C denotes the output of the stage, as shown in Figure 8.6.b [216]. *Stage$_i$* contains a token if its output C_i is not equal to the output C_{i+1} of *stage$_{i+1}$*. However, *stage$_i$* contains a bubble if its output C_i is equal to the output C_{i+1} of *stage$_{i+1}$*:

$$C_i \neq C_{i+1} \Longleftrightarrow Stage_i \longleftarrow Token$$
$$C_i = C_{i+1} \Longleftrightarrow Stage_i \longleftarrow Bubble$$

The number of tokens and bubbles will be respectively denoted N_T and N_B. According to the previous definitions, N_T must be an even number. Each stage of the ring contains either a token or a bubble. $N_T + N_B = N$, where N is the number of the ring stages. Suppose that there is a token in *stage$_i$*. This token will move to *stage$_{i+1}$*, if and only if *stage$_{i+1}$* contains a bubble. Self-timed ring produces two different modes of oscillation: Evenly Spaced or Burst modes. In the evenly spaced mode, the events inside the ring are equally spaced in time. In the burst mode the events are spaced in time in a non homogenous way [326].

TABLE 8.1

C-element truth table

a	b	y
0	0	0
0	1	y
1	0	y
1	1	1

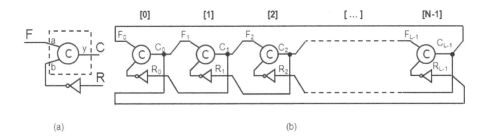

(a)   (b)

FIGURE 8.6

A self-timed ring.

8.2.2.4   Sensing the computational activity

As mentioned previously activity sensors play a critical role in DVFS systems and must be selected carefully. However, we are facing many difficulties in order to implement a sensor that measures the processor activity accurately. Different assumptions were proposed, here is an example for two of them:

Current Activity Sensor It consists of an analog solution to monitor the activity of the system with respect to the amount of current consumed. In [260], a monitor fabricated in a 90nm CMOS technology which is able to estimate the circuit activity is proposed. Unfortunately, this kind of sensors has very limited applications, for example in our case it's difficult to say if the activity is strong or not. Indeed, the current is data-dependant for a part and activity-dependant for the other part (we can say that the current gives an image of the executed instructions). Moreover, it is also difficult to plan the deadline of the task, so this sensor is not well-suited for our DVFS application.

Instruction Counter There exist two possible solutions to implement this kind of activity sensing:

- *Direct instruction count:* By incrementing the counter each time a new instruction has been executed and calculates the average value with respect to a reference clock. It's a simple and efficient way to estimate the activity and the progress in a task. The limit is that it is difficult to accurately compute the time. Indeed, with CISC machine (not RISC), it is difficult to predict the remaining clock cycles or the speed for reaching a deadline.

- *Statically computing the clock cycles:* At compile time, we calculate a bound for the number of clock cycles. We estimate here, instruction by instruction, the number of cycles. This is a little bit more complex and more accurate. The gain is not formally proven compared to a simple counter.

8.3 Controlling uncertainty and handling process variability

Embedded integrated systems have two means of implementation. Firstly, the conventional clocked circuits with their global synchronization in which we face the huge challenge of generating and distributing a low-skew global clock signal and reducing the clock tree power consumption of the whole chip. As a result, the global synchronization with large systems is difficult to implement. Secondly, SoCs, built with predesigned IP-blocks running at different frequencies, need to integrate all the IP-blocks into a single chip. Therefore, global synchronization tends to be impractical [97]. By removing the globally distributed clock, GALS circuits provide a promising solution for SoC design. Moreover, GALS techniques allow each locally synchronous modules frequency and voltage to be set independently, making scaling far more convenient than with the standard synchronous approach. It is possible to set the optimal frequency for a GALS module, because all interblock communication is performed asynchronously. GALS are chips divided into multiple frequency domains, where each domain is synchronous with respect to its clock. The different domains are mutually asynchronous in that they may run at different clock frequencies (see Figure 8.7).

A GALS architecture can mitigate the impact of process and temperature variations, because a globally asynchronous system does not require that the global frequency was dictated by the longest path delay (the critical path) of the whole chip. In
this case, each clock-domain frequency is only determined by the slowest path in its
domain. In a GALS system process variability and fabrication yield can be improved
by smartly removing tasks over fault or low performance nodes and assign them into
other high performance ones, as shown in Figure 8.8. As each processing node performance is measured by a sensor, a global system manager is able to distribute the
tasks over the nodes. The task assignment takes into account the node performances
and the task processing loads. The main target of this manager is to guarantee an overall chip performance. With this kind of approach, it is no more required to separately
guarantee a performance for each node which relaxes the fabrication constraints and
permits a yield enhancement.

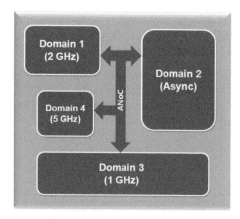

FIGURE 8.7
GALS architecture.

Based on the previous ideas we propose a solution to manage the impact of process variability in a multiprocessor GALS system. In a GALS system, we can choose
to slow down some parts of the circuit while allowing others to operate at the maximum frequency. This enables more energy saving opportunities than conventional
systems built around one processor and allows adapting the clock speed to the local process quality. Moreover, it has also been shown that multiple-clock designs
(GALS) with voltage scaling are even better not only in terms of power and performance but also in terms of variability [192]. As a result building a system based on
the implementation of hardware resources whose performances are unpredictable at
the fabrication time requires having total management strategies of the performance
by adaptation of the voltage/frequency in order to respect the real time constraints of
the application and the allocated energy budget. So it is proposed to use automatic
feedback loops based on:

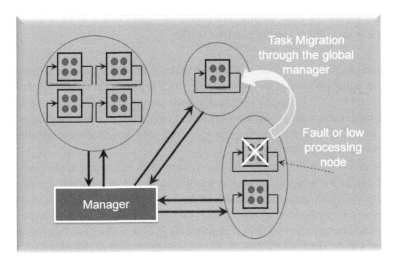

FIGURE 8.8
Fabrication yield control by distributing tasks over different processing nodes in a
GALS system.

1. Measurement of the real local performances of silicon and the actuation of the
   parameters voltage/frequency (hardware level)

2. The suitable hardware resource allocation for the execution of a task in the
   assigned time/energy budget (operating system level)

The idea is the use of dynamic DVFS with task scheduling techniques to dynamically manage not only the energy budget but also the activity of the processing node
based on advanced automatic techniques. These techniques will allow an optimal regulation of the frequency/voltage of the converter according to the computational load
and the load distribution in the various GALS processors. In order to compensate for
the process variation due to the technology dispersion, and optimize the operation of
the circuit, the dynamic voltage/frequency regulation itself should be self-adjustable
and robust against process variability with the variable loads and dispersion models.
Implementing DVFS in a GALS system presents many design challenges:

1. Specific mechanisms are needed, allowing the different domains to communicate and synchronize in a reliable manner.

2. Splitting the computational load on the processors must be done as a tradeoff
   with the communication load. Usually the communication systems consume a
   lot of energy.

3. Interdependencies between different clock domains impact the overall performances. A domain operating at a low speed tends to slow down the communications and to reduce the processor speed in communication.

4. Sensing the local process quality and processor speed can also be a design issue as it was discussed in the previous section.

Many researches have been done in this field to develop DVFS and to guarantee a correct system behavior even if the process variations are huge. In GALS systems [271], a model has been developed to gradually decrease the clock frequency when there are little observable changes in the workload. Compiler for controlling GALS systems have been developed too [191], but there are very few studies integrating power and process variation management.

One of the propositions to handle the uncertainty of a processing node over a GALS system due to the impact of process variability and also to reduce its energy-consumption by means of automatic control methods is the use of activity sensors. Activity sensors were embedded in each processing unit. These sensors provide a real performance measurement of different processing nodes after the fabrication process, which will be used afterwards by the operating system to distribute tasks over different processing nodes and assign those low or fault processing node tasks over idle ones. This means the need for the rescheduling of tasks in each processing node to meet the new assigned deadlines, and this will be achieved by controlling its voltage/frequency, which in accordance will control its energy of consumption. Figure 8.9 shows this control system using three different control loops. The control loops are applied in different architecture levels: control in energy of consumption (supply voltage), control in the processing power (supply voltage/clock frequency) and in the management of the Quality of Service (QoS) provided by the application. Voltage, frequency and energy control loops are used in order to adapt the energy of consumption and the process variability effect. The other control loop is needed to deal with to the QoS (at the application level), limitation of processing power and/or channel of communication and constraints in energy consumption.

The QoS controller sends a set of information (required speed, number of instructions, and deadlines for each instruction), which are given by the operating system (that can be statically inserted into the code or dynamically computed at run time by the OS). There are also sensors embedded in each processing unit in order to provide real-time measurements of the processor speed. Therefore, combining this information about the real-time requirements of the applications and the OS enables us to create a computational load profile with respect to time. Consequently, using such a profile makes possible to apply a fine-grain power/energy management allowing application deadlines to be met. Note that task scheduling is an important issue when using this technique. Sophisticated scheduling policies, as the one proposed in [336], can easily take advantage of this work by simply interfacing to the hardware features proposed using an API. Here, the DVFS hardware part contains a DC/DC converter for voltage regulation and a programmable clock generator for frequency regulation. The energy controller dynamically controls the DC/DC converter to scale the supply voltage as well as the clock frequency in order to satisfy the application computational needs with an appropriate management strategy. In this way we could manage the impact of process variability by the application of DVFS with task scheduling techniques, under the control of an Energy/Performance controller.

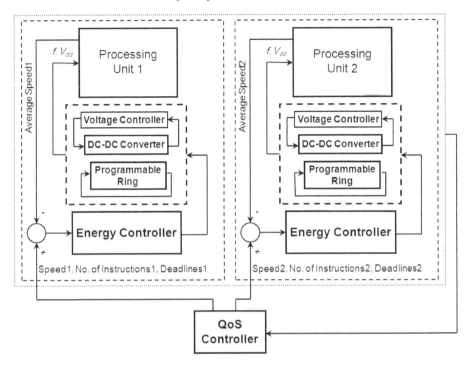

FIGURE 8.9
Energy/performance control simplified architecture.

8.4 Data synchronization in GALS system

In a GALS systems each synchronous part, usually referred as clock domain, operates with its own clock signal. Therefore, synchronization remains an issue either the clock domains work with different frequencies, or with the same central clock frequency. This is so because the communications among clock domains demand harmonization of the different clock phases in order to guarantee that data is reliably transferred among clock domains.

Several GALS methods address the problem of safe and reliable data transfer between independent clock domains. For example, Mullins and Moore give a detailed GALS analysis based on the clock generation processes and I/O port operations of the various methods [216]. Clock Domain Crossing (CDC) strategies are required to achieve this purpose. Several hardware techniques have been proposed to deal with this problem [70, 179, 235, 251]. An interesting taxonomy of CDC hardware techniques is presented in [70]. Three main strategies are outlined in [70], namely, pausible-clock generators, FIFO buffers, and boundary synchronization:

Pausible-clock generators: Applying local (pausible, stretchable, or data-driven) clocking, which avoids metastability by ensuring no clock pulses are generated when data is transferred.

FIFO buffers: Using asynchronous FIFO buffers between locally synchronous blocks to hide the synchronization problem. A SoC architecture that uses distinct clock domains connected through bisynchronous FIFO buffers is commonly called a GALS system. In our case, however, we refer only to pure GALS systems, in which the blocks are connected asynchronously.

Boundary synchronization: Performing boundary synchronization on the signals crossing the borders of the locally synchronous island without stopping the complete locally synchronous block during data transfer.

8.4.1   GALS wrapper with pausible clocking

Many GALS systems presented in the past few years use pausible (or stretchable) clocking [224, 326, 331]. The basic idea of all these proposals is similar: transferring data between wrappers when both the data transmitter and data receiver clocks are stopped. This elegantly solves the problem of synchronization between the two clock domains. Figure 8.10 illustrates the general structure of such a system. The asynchronous wrapper contains input and output ports that perform the handshake process between the locally synchronous modules, and it generates a stretch signal to stop the activity of both clocks. The basic GALS method focuses on point-to-point communication between blocks.

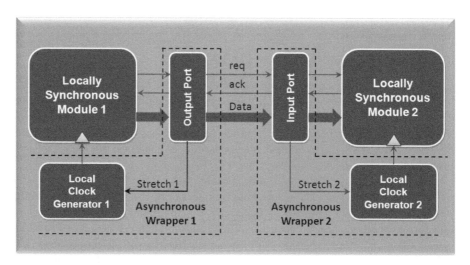

FIGURE 8.10
Globally asynchronous, locally synchronous (GALS) system with pausible clocking.

8.4.2 FIFO solutions

Another approach for interfacing locally synchronous blocks is using specially designed asynchronous FIFO buffers and hiding the system synchronization problem within the FIFO buffers [25, 55, 56]. Such a system can tolerate very large interconnect delays and is also robust with regard to metastability. Designers can use this method to interconnect asynchronous and synchronous systems and also to construct synchronous-synchronous and asynchronous-asynchronous interfaces. Figure 8.11 diagrams a typical FIFO interface, which achieves an acceptable data throughput [56]. In addition to the data cells, the FIFO structure includes an empty/full detector and a special deadlock detector.

The advantage of FIFO synchronizers is that they do not affect the locally synchronous modules operation. However, with very wide interconnect data buses, FIFO structures can be costly in silicon area. Also, they require specialized complex cells to generate the empty/full flags used for flow control. The introduced latency might be significant and unacceptable for high-speed applications. As an alternative, E. Beigne and P. Vivet designed a synchronous-asynchronous FIFO based on the bisynchronous classical FIFO design using gray code, for the specific case of an asynchronous NoC interface [25]. Their aim was to maintain compatibility with existing design solutions and to use standard CAD tools. Thus, even with some performance degradation or suboptimal architecture, designers can achieve the main goal of designing GALS systems in the standard design environment.

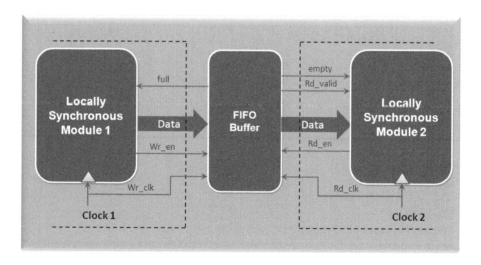

FIGURE 8.11
Typical FIFO-based GALS system.

8.4.3 Boundary synchronization

A third solution is to perform data synchronization at the borders of the locally synchronous island, without affecting the inner operation of locally synchronous blocks and without relying on FIFO buffers. For this purpose, designers can use standard two-flop, one-flop, predictive, or adaptive synchronizers for mesochronous systems, or locally delayed latching [81, 107]. This method can achieve very reliable data transfer between locally synchronous blocks. On the other hand, such solutions generally increase latency and reduce data throughput, resulting in limited applicability for high-speed systems.

In Figure 8.12 a practical example of a boundary synchronized GALS based system is shown. It uses the advantage of asynchronous logic in securing data communication between different clock domains, while it keeps avoiding the problem of metastability. it uses a Self-Timed Ring and a kind of control switch that synchronizes the transmission and reception of data using asynchronous Network-on-Chip (NoC) interface (i.e., request and acknowledgement signals). The main advantage of this technique is the avoidance of stopping data transfer between different clock domains during their synchronization.

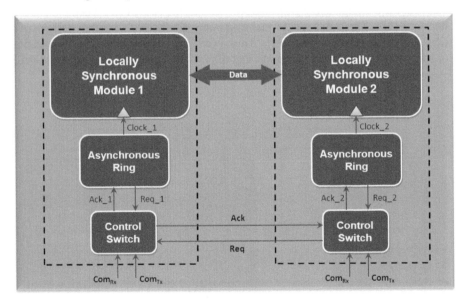

FIGURE 8.12
Boundary synchronization-based GALS system.

Figure 8.13 shows a timing diagram explaining an example of the boundary synchronized method. We have two case studies:

Case (1) Here *Module 1* operates at a frequency lower than that of *Module 2*. It is shown that with no communication where Com_{Tx} and Com_{Rx} of both modules are 0, the two module clock inputs $Clock_1$ and $Clock_2$ runs freely. When we have a communication through the asynchronous NoC interface, for example a data transmission from *Module 1* where $Com1_{Tx} = 1$ and $Com1_{Rx} = 0$ to *Module 2* where $Com2_{Rx} = 1$ and $Com2_{Tx} = 0$, $Clock_2$ input synchronizes with $Clock_1$ during the communication period where $Com1_{Tx}$ and $Com2_{Rx} = 1$, as $Clock_1$ is the slowest operating frequency of the two synchronous modules. So now the two modules are working at the same clock frequency, with no metastability problem. Once this communication between the two synchronous modules is disconnected, each module returns again to its own operating frequency.

Case (2) Here we have a data reception by *Module 1* where $Com1_{Rx} = 1$ and $Com1_{Tx} = 0$ from *Module 2* where $Com2_{Tx} = 1$ and $Com2_{Rx} = 0$, with *Module 1* still operating at a frequency lower than that of *Module 2*. Even that *Module 1* is now working as a receiver, again $Clock_1$ will not change, as it's still the slowest frequency of two communicating modules. On the other hand $Clock_2$ of *Module 2* which is now the transmitter, will synchronize with $Clock_1$ of *Module 1* during the communication period where $Com1_{Rx}$ and $Com2_{Tx} = 1$. So now the two modules are working again at the same clock frequency, with no metastability problem, and once this communication is disconnected, each module will return to its own operating frequency.

As a conclusion, once the clock frequency of a communicating module (transmitter or receiver) is slower than that of the other communicating one, it will continue to work at the same clock frequency level. On the other hand, the faster communicating module (transmitter or receiver) will always synchronize its clock frequency with that of the slower communicating one. In this way it's always guaranteed whenever we have a communication over the asynchronous NoC, the two communicating domains are working with the slowest clock frequency of both synchronous modules to avoid any mis-synchronization problem.

Table 8.2 summarizes the properties of GALS systems synchronization methods. Contrary to earlier expectations, GALS-based solutions dont automatically offer performance gains. Interblock communication incurs some penalty in all GALS systems. In pausible-clock systems, the clock can be stretched when transferring data on slow communication links, reducing the locally synchronous modules operating frequency. FIFO-based systems, depending on the communication link, suffer from additional latency. If designed carefully, performance degradation in a GALS system will be insignificant; however, in some examples (for various reasons), the reported performance degradation of the GALS system was as high as 23% [224].

The GALS approach is a vehicle for block interconnects. A crucial parameter for such an application is data throughput and latency. For many GALS solutions, the problem of data throughput is critical. Some pausible-clocking schemes can theoretically reach a maximum data throughput of one data item per clock cycle [224]. However, more often, data transfers are limited to every second clock cycle or even every fourth or fifth clock cycle of the locally synchronous block. In addition, in an environment with intensive data transfers the performance degrade significantly. For FIFO-based solutions, the throughput problem is less severe, but latency increases.

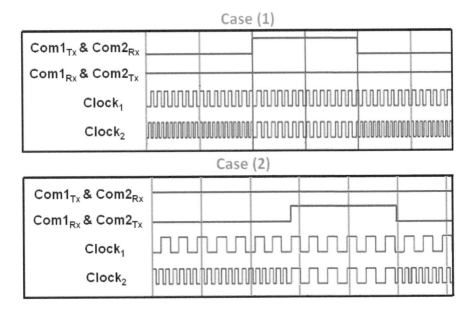

FIGURE 8.13
Timing diagram for a boundary synchronized system.

8.5 Conclusions

Regarding the recent and upcoming CMOS technologies, designing a complex SoC is now becoming very challenging. These challenges are due to several constraints, such as guaranteeing the correct system behavior with strong process variations or controlling the system speed and energy. The main problem with such systems is that their modeling is complex and their behavior is relatively difficult to predict. As they integrate several interdependent control mechanisms, SoCs require many competences in microelectronic design, in software and, more recently, in complex and networked control systems.

This chapter presents a survey of different problems facing designers over the nanometric era. An analysis and a solution to each of these challenges are presented, which enables us to design a self-adaptable SoC. GALS systems were considered, but the proposed solutions are also applicable to simpler designs. For example, DVFS is applied to make self-adaptable SoC to its constraints in term of energy. In order to be tolerant to the process variability, specific sensors were also proposed to evaluate the fabrication process quality and the local environmental parameters (voltage, temperature) in each clock domain. The outputs of these sensors, smartly combined with the DVFS, provide the required information to determine the appropriate clock frequencies which do not violate the local timing constraint. Therefore, the SoC also is

TABLE 8.2
Properties of GALS techniques

Property	Synchronization Method		
	Pausible Clocking [162,224,331]	**FIFO-based** [25,55,56]	**Boundary Synchronization** [37,81]
Area Overhead	Low	Medium to high	Low
Latency	Low	High	Medium
Throughput	Lowered according to clock pause rate	High	Medium
Power Consumption	Low	High	Medium
Additional Cells	Mutex, delay-line, Muller-C	Empty/full flag	Muller-C, mutex
Advantages	No metastability	Simple solution, throughput	Low overhead
Disadvantages	Local clock generators, throughput	Area overhead, latency	Requires verification, throughput

able to self-adapt its speed and voltage to the process variations. This implies the use of feedback systems and the implementation of a well-suited control to efficiently manage the process variability as well as the energy or the speed. This leads to a potential fabrication yield improvement because the system is able to adapt itself to the process variation in order to guarantee the correctness of its overall behavior. This implies the use of GALS systems and the implementation of an efficient NoC. Since data synchronization over NoCs used in GALS is mandatory to avoid metastability problems, a comparison among different techniques of synchronization is discussed.

Today, the trends in SoC design show clearly the path and there is no doubt that many of these techniques are and will be integrated in SoCs. This is required for relaxing the design constraints (e.g., timing, power, variability) and for facilitating the reuse of IP blocks which are connected only to an NoC (no more timing closure issues). This leads to an enhancement of the design productivity and an improvement of the fabrication yield, thanks to the chip self-adaptability.

8.6 Glossary

AMS: Analog and Mixed Signal

ANoC: Asynchronous Network-on-Chip

API: Application Programming Interface

CDC: Clock Domain Crossing

CISC: Complex Instruction Set Computer

DFS: Dynamic Frequency Scaling

DVFS: Dynamic Voltage and Frequency Scaling

DVS: Dynamic Voltage Scaling

FIFO: First in First out buffer

GALS: Globally Asynchronous Locally Synchronous

HVT: High Voltage Threshold

LVT: Low Voltage Threshold

MIPS: Millions Instructions per Second

NoC: Network-on-Chip

OS: Operating System.

PID: Proportional Integral Derivative Controller

PLL: Phase Locked Loop

QoS: Quality of Service

RISC: Reduced Instruction Set Computer

SoC: System on Chip

SVT: Standard Voltage Threshold

VCO: Voltage Controlled Oscillator

Bibliography

[1] RODIN: Rigorous Open Development Environment for Complex Systems. http://rodin.cs.ncl.ac.uk/.

[2] J. R. Abrial, editor. *The B-Book: Assigning Programs to Meanings.* Cambridge University Press, 1996.

[3] J. Adamek, H. Herrlich, and G. Strecker. *Abstract and Concrete Categories.* Dover Publications, 2009.

[4] C. Addo-Quaye. Thermal-Aware Mapping and Placement for 3-D NoC Designs. In *Proceedings of the IEEE International System-on-Chip Conference*, pages 25–28, 2005.

[5] B. Ahmad, A. T. Erdogan, and S. Khawam. Architecture of a Dynamically Reconfigurable NoC for Adaptive Reconfigurable MPSoC. In *Proceedings of the 1st NASA/ESA Conference on Adaptive Hardware and Systems (AHS'06)*, pages 405–411, 2006.

[6] E. Anceaume, X. Défago, M. Gradinariu, and M. Roy. Towards a Theory of Self-Organization. In *Principles of Distributed Systems*, volume 3974 of *Lecture Notes in Computer Science*, pages 191–205. Springer-Verlag, 2006.

[7] J.H. Anderson, G. Prencipe, and R. Wattenhofer, editors. *Principles of Distributed Systems*, volume 3974 of *Lecture Notes in Computer Science*. Springer-Verlag, 2006.

[8] G. Ascia, V. Catania, and M. Palesi. Mapping Cores on Network-on-Chip. *International Journal of Computational Intelligence Research*, 1(2):109–126, December 2005.

[9] A. Asperti and G. Longo. *Categories, Types and Structures.* MIT Press, 1991.

[10] O. Babaoglu, G. Canright, A. Deutsch, G. A. D. Caro, F. Ducatelle, L. M. Gambardella, et al. Design Patterns from Biology for Distributed Computing. *ACM Transactions on Autonomous and Adaptive Systems (TAAS)*, 1(1):26–66, 2006.

[11] R. J. Back and R. Kurki-Suonio. Decentralization of Process Nets with Centralized Control. In *Proceedings of the 2nd ACM SIGACT-SIGOPS Symposium on Principles of Distributed Computing*, pages 131–142, 1983.

[12] R. J. Back and K. Sere. From Action Systems to Modular Systems. *Software – Concepts and Tools*, 17:26–39, 1996.

[13] R. J. Back and K. Sere. Superposition Refinement of Reactive Systems. *Formal Aspects of Computing*, 8(3):324–346, 1996.

[14] C. Badica, G. Mangioni, V. Carchiolo, and D. D. Burdescu, editors. *Intelligent Distributed Computing, Systems and Applications*, volume 162 of *Studies in Computational Intelligence*. Springer-Verlag, 2008.

[15] C. Badica and M. Paprzycki, editors. *Advances in Intelligent and Distributed Computing*, volume 78 of *Studies in Computational Intelligence*. Springer-Verlag, 2008.

[16] J. C. M. Baeten and W. P. Weijland. *Process Algebra*. Cambridge University Press, October 1990.

[17] M. Bakhouya. Evaluating the Energy Consumption and the Silicon Area of On-chip Interconnect Architectures. *Journal of Systems Architecture*, 55(7-9):387–395, 2009.

[18] M. Bakhouya. Towards a Bio-Inspired Architecture for Autonomic Network-on-Chip. In *Proceedings of the Workshop on Autonomic and High Performance Computing (AHPC'10)*, pages 491–497, 2010.

[19] J. Balfour and W. J. Dally. Design Tradeoffs for Tiled CMP On-Chip Networks. In *Proceedings of the 20th Annual International Conference on Supercomputing (ICS'06)*, pages 187–198, 2006.

[20] N. Banerjee, P. Vellanki, and K. S. Chatha. A Power and Performance Model for Network-on-Chip Architectures. In *Proceedings of the Conference on Design, Automation and Test in Europe Conference and Exhibition*, volume 2, pages 1250–1255, 2004.

[21] M. Banikazeni and P. K. Dhabaleskwar. *Efficient Scatter Communication in Wormhole k-Ary n-Cubes with Multidestination Message Passing*. Technical Report OSU-CISRC-9/96-TR46, 1996.

[22] A. Bar-Noy and S. Kipnis. Designing Broadcasting Algorithms in the Postal Model for Message-Passing Systems. In *Proceedings of the 1992 Symposium on Parallel Algorithms and Architectures*, pages 13–22. ACM, 1992.

[23] M. Barnett, L. Shuler, R. van de Geijn, S. Gupta, D.G. Payne, and J. Watts. Interprocessor Collective Communication Library (InterCom). In *Proceedings of the Scalable High Performance Computing Conference*, pages 357–364. IEEE Computer Society, 1994.

[24] J. Becker, K. Brändle, U. Brinkschulte, J. Henkel, W. Karl, T. Köster, et al. Digital On-Demand Computing Organism for Real-Time Systems. In *Proceedings of ARCS Workshops*, pages 230–245, 2006.

[25] E. Beigne and P. Vivet. Design of On-Chip and Off-Chip Interfaces for a GALS NoC Architecture. In *Proceedings of the 12th IEEE International Symposium on Asynchronous Circuits and Systems (ASYNC'06)*, pages 172–183. IEEE Computer Society, 2006.

[26] S. Bell, B. Edwards, J. Amann, R. Conlin, K. Joyce, V. Leung, et al. TILE64 - Processor: A 64-Core SoC with Mesh Interconnect. In *Digest of Technical Papers. IEEE ISSCC 2008*, pages 88–598, 2008.

[27] L. Benini. Application Specific NoC Design. In *Proceedings of the IEEE Design, Automation and Test in Europe Conference (DATE'06)*, volume 1, pages 1–5, Munich, Germany, March 6–10, 2006.

[28] L. Benini, A. Macii, M. Poncino, and R. Scarsi. Architectures and Synthesis Algorithms for Power-Efficient Bus Interfaces. *IEEE Transactions on CAD*, 19(9):969–980, 2000.

[29] L. Benini and G. De Micheli. Networks on Chips: A New SoC Paradigm. *IEEE Computer*, 35(1):70–78, January 2002.

[30] L. Benini and G. De Micheli. *Networks on Chips: Technology and Tools*. Morgan Kaufmann, 2006.

[31] G. M. Bergman. *An Invitation to General Algebra and Universal Constructions*. Henry Helson, 1998.

[32] M. Bernardo and A. Cimatti, editors. *Formal Methods for Hardware Verification*, volume 3965 of *Lecture Notes in Computer Science*. Springer-Verlag, 6th International School on Formal Methods for the Design of Computer, Communication, and Software Systems (SFM 2006), Bertinoro, Italy, Advanced Lectures, May 22-27 2006.

[33] K. Bernstein, D. J. Frank, A. E. Gattiker, W. Haensch, B. L. Ji, S. R. Nassif, et al. High-Performance CMOS Variability in the 65-nm Regime and Beyond. *IBM Journal of Research and Development*, 50(4/5):433–449, 2006.

[34] F. Berthelot, F. Nouvel, and D. Houzet. A Flexible System Level Design Methodology Targeting Run-Time Reconfigurable FPGAs. *EURASIP Journal on Embedded Systems*, 2008.

[35] R. H. Bisseling. *Parallel Scientific Computation: A Structured Approach Using BSP and MPI*. Oxford University Press, 2004.

[36] T. Bjerregaard and K. Mahadevan. A Survey of Research and Practices of Network-on-Chip. *ACM Computing Surveys*, 38(1):1–51, March 2006.

[37] T. Bjerregaard, S. Mahadevan, R. G. Olsen, and J. Sparsø. An OCP Compliant Network Adapter for GALS-Based SoC Design Using the MANGO Network-on-Chip. In *Proceedings of the International Symposium on System-on-Chip (SOC'05)*, pages 171–174, 2005.

[38] C. Bobda, A. Ahmadinia, M. Majer, J. Teich, S. Fekete, and J. van der Veen. DyNoC: A Dynamic Infrastructure for Communication in Dynamically Reconfigurable Devices. In *Proceedings of the International Conference on Field Programmable Logic and Applications*, pages 153–158, 2005.

[39] E. Bolotinand, I. Cidon, R. Ginosar, and A. Kolodny. QNoC: QoS Architecture and Design Process for Network on Chip. *Journal of Systems Architecture*, 50(2–3):105–128, 2004.

[40] E. Bonabeau, M. Dorigo, and G. Theraulaz. *Swarm Intelligence: From Natural to Artificial Systems*. Oxford University Press, August 1999.

[41] L. Bononi and N. Concer. Simulation and Analysis of Network on Chip Architectures: Ring, Spidergon and 2D Mesh. In *Proceedings of the IEEE Design, Automation and Test in Europe Conference (DATE'06)*, volume 2, pages 154–159, Munich, Germany, March 6–10, 2006.

[42] L. Booker. Genetic Algorithms and Simulated Annealing. In *IEEE Computer*. Morgan Kaufmann, 1987.

[43] S. Borkar, N. P. Jouppi, and P. Stenstrom. Microprocessors in the Era of Terascale Integration. In *Proceedings of the Conference on Design, Automation and Test in Europe (DATE'07)*, pages 237–242. EDA Consortium, 2007.

[44] D. Borrione, A. Helmy, L. Pierre, and J. Schmaltz. A Formal Approach to the Verification of Networks on Chip. *EURASIP Journal of Emb. Sys.*, 2009.

[45] B. Bose, B. Broeg, Y. Known, and Y. Ashir. Lee Distance and Opological Properties of k-Ary n-Cubes. *IEEE Transactions on Computers*, 44(8):1021–1030, 1995.

[46] A. Bouillard, B. Gaujal, S. Lagrange, and E. Thierry. Optimal Routing for End-to-End Guarantees Using Network Calculus, Performance Evaluation. *Performance Evaluation*, 64(11–12):883–906, 2008.

[47] F. R. Boyer, H. G. Epassa, and Y. Savaria. Embedded Power-Aware Cycle by Cycle Variable Speed Processor. In *Proceedings of the IEE conference on Computers and Digital Techniques*, volume 153, pages 283–290, July 2006.

[48] A. Brindle. *Genetic Algorithms for Function Optimization*. PhD thesis, University of Alberta–Edmondton, 1981.

[49] C. J. Burgess and A. G. Chalmers. Genetic Algorithms for Generating Minimum Path Configurations. *Microprocessors and Microsystems*, 19(1), 1995.

[50] C. J. Burgess and A. G. Chalmers. *Optimum Transputer Configurations for Real Applications Requiring Global Communications*. Technical Report UMI Order Number: CS-EXT-1995-044, University of Bristol, 1995.

[51] T. Canhao, A. W. Yin, P. Liljeberg, and H. Tenhunen. A Study of 3D Network-on-Chip Design for Data Parallel H.264 Coding, Submitted 2010.

[52] P. Carlone, G. S. Palazzo, and R. Pasquino. Pultrusion Manufacturing Process Development: Cure Optimization by Hybrid Computational Methods. *International Journal of Computers and Mathematics with Applications, Elsevier*, 53(9):1464–1471, April 2007.

[53] G. Di Caro and M. Dorigo. AntNet: Distributed Stigmergetic Control for Communications Networks. *Journal of Artificial Intelligence Research*, 9:317–365, 1998.

[54] E. Carvalho, N. Calazans, and F. Moraes. Heuristics for Dynamic Task Mapping in NoC-Based Heterogeneous MPSoCs. In *Proceedings of the 18th IEEE International Workshop on Rapid System Prototyping*, pages 4–40, 2007.

[55] A. Chakraborty and C. A. Chakraborty. Efficient Self-Timed Interfaces for Crossing Clock Domains. In *Proceedings of the 9th International Symposium on Asynchronous Circuits and Systems*, pages 78–88, 2003.

[56] T. Chelcea and S. M. Nowick. Low-Latency Asynchronous FIFO's Using Token Rings. In *Proceedings of the 6th International Symposium on Advanced Research in Asynchronous Circuits and Systems (ASYNC'00)*, pages 210–220. IEEE Computer Society, 2000.

[57] G. Chelius and E. Fleury. NP-Completeness of Ad Hoc Multicast Routing Problems. Research Report 5665, Institut National de Recherche en Informatique and eten Automatique, 2005.

[58] G. Chen, F. Li, S. W. Son, and M. Kandemir. Application Mapping for Chip Multiprocessors. In *Proceedings of the 45th IEEE/ACM/EDA Design Automation Conference*, pages 620–625, 2008.

[59] X. Chen and L. Peh. Leakage Power Modeling and Optimization in Interconnection Networks. In *Proceedings of the International Symposium on Low Power Electronics and Design (ISLPED'03)*, pages 90–95, 2003.

[60] R. Cheng, M. Gen, and Y. Tsujimura. A Tutorial Survey of Job-Shop Scheduling Problems Using Genetic Algorithms: I. Representation. *Computer and Industrial Engineering*, 35(2), 1999.

[61] G. M. Chiu. The Odd-Even Turn Model for Adaptive Routing. *IEEE Transactions on Parallel and Distributed Systems*, 11(7):729–738, 2000.

[62] M. H. Cho, M. Lis, K. S. Shim, M. Kinsy, T. Wen, and S. Devadas. Oblivious Routing in On-Chip Bandwidth-Adaptive Networks. In *Proceedings of the 18th International Conference on Parallel Architectures and Compilation Techniques*, pages 181–190, 2009.

[63] C. L. Chou and R. Marculescu. Contention-Aware Application Mapping for Network-on-Chip Communication Architectures. In *Proceedings of the IEEE International Conference on Computer Design*, pages 164–169, 2008.

[64] C. L. Chou and R. Marculescu. User-Aware Dynamic Task Allocation in Networks-on-Chip. In *Proceedings of the Conference on Design, Automation and Test in Europe (DATE '08)*, pages 1232–1237, 2008.

[65] C. L. Chou, U. Y. Ogras, and R. Marculescu. Energy- and Performance-Aware Incremental Mapping for Networks on Chip with Multiple Voltage Levels. *IEEE Transactions on Computer-Aided Design of Integrated Circuits and Systems*, 27(10), 2008.

[66] C. Ciordas, T. Basten, A. Radulescu, and K. Goossens. An Event-Based Network-on-Chip Monitoring Service. In *Proceedings of the 9th IEEE International High-Level Design Validation and Test Workshop (HLDVT'04)*, pages 149–154, Sonoma Valley, CA, November 10–12, 2004.

[67] C. Clos. A Study of Non-Blocking Switching Networks. *Bell System Technical Journal*, 32(9):406–424, 1953.

[68] M. Coenen, S. Murali, A. Radulescu, K. Goossens, and G. De Micheli. A Buffer-Sizing Algorithm for Networks-on-Chip Using TDMA and Credit-Based End-to-End Flow Control. In *Proceedings of the 4th International Conference on Hardware/Software Codesign and System Synthesis (CODES+ISSS'06)*, pages 130–135, 2006.

[69] I. R. Cohen. Real and Artificial Immune Systems: Computing the State of the Body. *Nature Reviews, Immunology*, 7:569–574, 2007.

[70] C. E. Cummings. *Clock Domain Crossing (CDC) Design and Verification Techniques Using System Verilog*. SNUG-2008, Boston, 2009.

[71] W. J. Dally. Virtual-Channel Flow Control. In *Proceedings of the 17th Annual International Symposium on Computer Architecture (ISCA'90)*, pages 60–68, 1990.

[72] W. J. Dally and B. Towles. Route Packets, not Wires: On-Chip Interconnection Networks. In *Proceedings of the 38th Design Automation Conference (DAC'01)*, pages 683–689, June 2001.

[73] W. J. Dally and B. Towles. *Principles and Practices of Interconnection Networks*. Computer Architecture and Design. Morgan Kaufmann, 2004.

[74] M. Daneshtalab, A. Sobhani, A. Afzali-Kusha, O. Fatemi, and Z. Navabi. NoC Hot Spot Minimization Using AntNet Dynamic Routing Algorithm. In *Proceedings of the IEEE 17th International Conference on Application-Specific Systems, Architectures and Processors (ASAP'06)*, pages 33–38, 2006.

[75] S. Das, C. Tokunaga, S. Pant, W. H. Ma, S. Kalaiselvan, K. Lai, et al. RazorII: In Situ Error Detection and Correction for PVT and SER Tolerance. *IEEE JSSC*, 44(1):32–48, 2009.

[76] D. Dasgupta, Z. Ji, and F. González. Artificial Immune Systems (AIS) Research in the Last Five Years. In *Proceedings of the Congress on Evolutionary Computation (CEC'03)*, pages 528–535, 2003.

[77] S. Dhar, E. Dhar, and D. Maksimovic. Switching Regulator with Dynamically Adjustable Supply Voltage for Low Power VLSI. In *Proceedings of the International Symposium on Low Power Electronics and Design*, pages 103–107, 2001.

[78] S. Dhar, D. Maksimović, and B. Kranzen. Closed-Loop Adaptive Voltage Scaling Controller for Standard-Cell ASICs. In *Proceedings of the International Symposium on Low Power Electronics and Design (ISLPED'02)*, pages 103–107. ACM, 2002.

[79] E. W. Dijkstra. A Note on Two Problems in Connetion with Graphs. *Numerische Mathematik*, 1(1):269–271, December 1959.

[80] Y. Ding, M. Kandemir, M. J. Irwin, and P. Raghavan. Adapting Application Mapping to Systematic Within-Die Process Variations on Chip Multiprocessors. In *High Performance Embedded Architectures and Compilers*, volume 5409 of *Lecture Notes in Computer Science*, pages 231–247. Springer-Verlag, 2009.

[81] R. Dobkin, R. Ginosar, and C.P. Sotiriou. Data Synchronization Issues in GALS SoCs. In *Proceedings of the 10th International Symposium on Asynchronous Circuits and Systems*, pages 170–179, 2004.

[82] J. Duato and S. Yalamanchili. *Interconnection Networks: An Engineering Approach*. Morgan Kaufman Publishers, Elsevier Science, 2003.

[83] T. Dumitras, S. Kerner, and R. Marculescu. Towards on-Chip Fault-Tolerant Communication. In *Proceedings of the Asia and South Pacific Design Automation Conference (ASP-DAC'03)*, 2003.

[84] T. Dumitras and R. Marculescu. On-Chip Stochastic Communication [SoC Applications]. In *Proceedings of Design, Automation and Test in Europe Conference and Exhibition (DATE'03)*, pages 790–795, 2003.

[85] V. Dumitriu and G. Khan. Throughput-Oriented NoC Topology Generation and Analysis for High Performance SoCs. *IEEE Transactions on Very Large Scale Integration (VLSI) Systems*, 17(10):1433–1446, October 2009.

[86] S. Edwards, L. Lavagno, E. A. Lee, and A. Sangiovanni-Vincentelli. Design of Embedded Systems: Formal Models, Validation, and Synthesis. *Proceedings of the IEEE*, 85(3):366–390, 1997.

[87] A. Ejlali, B. Al Hashimi, P. Rosinger, and S. Miremadi. Joint Consideration of Fault-Tolerance, Energy-Efficiency and Performance in on-Chip Networks. In *Proceedings of the IEEE Design, Automation and Test in Europe Conference and Exhibition (DATE'07)*, pages 1647–1652, Nice, France, April 16–20, 2007.

[88] H. Elmiligi, A. A. Morgan, M. W. El-Kharashi, and F. Gebali. Power-Aware Topology Optimization for Networks-on-Chips. In *Proceedings of the IEEE International Symposium on Circuits and Systems (ISCAS'08)*, pages 360–363, Seattle, WA, USA, May 18–21, 2008.

[89] H. Elmiligi, A. A. Morgan, M. W. El-Kharashi, and F. Gebali. A Delay-Aware Topology-Based Design for Networks-on-Chip Applications. In *Proceedings of the 4th International Design and Test Workshop (IDT'09)*, pages 1–5, Riyadh, Saudi Arabia, November 15–17, 2009.

[90] H. Elmiligi, A. A. Morgan, M. W. El-Kharashi, and F. Gebali. A Reliability-Aware Design Methodology for Networks-on-Chip Applications. In *Proceedings of the 4th IEEE International Conference on Design and Technology of Integrated Systems in Nanoscale Era (DTIS'09)*, pages 107–112, Cairo, Egypt, April 6–9, 2009.

[91] H. Elmiligi, A. A. Morgan, M. W. El-Kharashi, and F. Gebali. Power Optimization for Application-Specific Networks-on-Chips: A Topology-Based Approach. *Journal of Microprocessors and Microsystems, Elsevier*, 33(5-6):343–355, August 2009.

[92] S. Fairbanks. Analog Micropipeline Rings for High Precision Timing. In *ASYNC'04: Proceedings of the 10th International Symposium on Asynchronous Circuits and Systems*, pages 41–50. IEEE Computer Society, 2004.

[93] M. A. Al Faruque, T. Ebi, , and J. Henkel. Run-Time Adaptive on-Chip Communication Scheme. In *Proceedings of the IEEE/ACM Int'l. Conf. on Computer-Aided Design (ICCAD'07)*, pages 26–31, 2007.

[94] M. A. Al Faruque, T. Ebi, and J. Henkel. Configurable Links for Runtime Adaptive on-Chip Communication. In *Proceedings of the Conference on Design, Automation and Test in Europe (DATE'09)*, pages 543–548, 2009.

[95] M. A. Al Faruque, R. Krist, and J. Henkel. ADAM: Run-Time Agent-Based Distributed Application Mapping for on-Chip Communication. In *Proceedings of the 45th IEEE/ACM/EDA Design Automation Conference (DAC'08)*, pages 760–765, 2008.

[96] S. P. Fekete, J. C. van der Veen, J. Angermeier, D. Göhringer, M. Majer, and J. Teich. Scheduling and Communication-Aware Mapping of HW/SW Modules for Dynamic Partial Reconfigurable SoC Architectures. In *Proceedings of the 20th International Conference on Architecture of Computer Systems*, 2007.

[97] L. Fesquet and H. Zakaria. Controlling Energy and Process Variability in System-on-Chips: Needs for Control Theory. In *Proceedings of the 3rd IEEE Multi-conference on Systems and Control (MSC'09)*, pages 302–307. IEEE Computer Society, 2009.

[98] D. Fick, A. DeOrio, G. Chen, V. Bertacco, D. Sylvester, and D. Blaauw. A Highly Resilient Routing Algorithm for Fault-Tolerant NoCs. In *Proceedings of Design, Automation Test in Europe Conference Exhibition (DATE'09)*, pages 21–26, 2009.

[99] K. Flautner, D. Flynn, D. Roberts, and D. I. Patel. IEM926: An Energy Efficient SoC with Dynamic Voltage Scaling. *Design, Automation and Test in Europe Conference and Exhibition*, 3:324–327, 2004.

[100] S. Franklin and A. Graesser. Is It an Agent, or Just a Program?: A Taxonomy for Autonomous Agents. In *Proceedings of the Workshop on Intelligent Agents III, Agent Theories, Architectures, and Languages (ECAI'96)*, pages 21–35. Springer-Verlag, 1997.

[101] J. Gaber. Spontaneous Emergence Model for Pervasive Environments. In *Proceedings of Globecom Workshops*, pages 1–4, November 2007.

[102] J. Gaber and M. Bakhouya. An Affinity-Driven Clustering Approach for Service Discovery and Composition for Pervasive Computing. In *Proceedings of IEEE ICPS'06*, pages 277–280, 2006.

[103] E. Gabrielyan and R. D. Hersch. Efficient Liquid Schedule Search Strategies for Collective Communications. In *Proceedings of the 12th IEEE International Conference on Network (ICON'04)*, volume 32, pages 760–766. Singapore, 2004.

[104] F. Gebali, H. Elmiligi, and M.W. El-Kharashi. *Networks-on-Chips: Theory and Practice*. CRC Press, Inc., 2009.

[105] D. Geer. Networks on Processors Improve On-Chip Communications. *IEEE Computer*, pages 17–20, March 2009.

[106] S. Gill. Parallel Programming. *Computer Journal*, 1:2–10, 1958.

[107] R. Ginosar. Fourteen Ways to Fool your Synchronizer. In *Proceedings of the 9th International Symposium on Asynchronous Circuits and Systems*, pages 89–96, 2003.

[108] D. E. Goldberg. *Genetic Algorithms in Search, Optimization, and Machine Learning*. Addison-Wesley, January 1989.

[109] C. Gomez, M. E. Gomez, P. Lopez, and J. Duato. Reducing Packet Dropping in a Bufferless NoC. *Euro-Par 2008: Parallel Processing*, 5168:899–909, 2008.

[110] K. Goossens. Formal Methods for Networks on Chips. In *Proceedings of the International Conference on Application of Concurrency to System Design (ACSD'05)*, pages 188–189, June 2005.

[111] K. Goossens, J. Dielissen, O. P. Gangwal, S. G. Pestana, A. Radulescu, and E. Rijpkema. A Design Flow for Application-Specific Networks on Chip with Guaranteed Performance to Accelerate SOC Design and Verification. In *Proceedings of the DATE Conference*, pages 1182–1187, 2005.

[112] G. W. T. Gordon and P. J. Bently. On Evolvable Hardware. *Soft Computing in Industrial Electronics*, pages 279–323, Physica-Verlag, 2002.

[113] P. Gratz, B. Grot, and S. W. Keckler. Regional Congestion Awareness for Load Balance in Networks-on-Chip. In *Proceedings of the 14th International Symposium on High-Performance Computer Architecture*, pages 203–214, 2008.

[114] D. Groth and S. Toby. *Network+ Study Guide*. Sybex, Inc., 2005.

[115] L. Guang, E. Nigussie, J. Isoaho, P. Rantala, and H. Tenhunen. Interconnection Alternatives for Hierarchical Monitoring Communication in Parallel SoCs. *Microprocessors and Microsystems*, 34(5):118–128, August 2010.

[116] L. Guang, E. Nigussie, P. Rantala, J. Isoaho, and H. Tenhunen. Hierarchical Agent Monitoring Design Approach towards Self-Aware Systems. *ACM Transactions on Embedded Computing Systems (TECS)*, 9(3):1–24, February 2010.

[117] G. Guindani, C. Reinbrecht, T. Raupp, N. Calazans, and F. G. Moraes. NoC Power Estimation at the RTL Abstraction Level. In *Proceedings of the IEEE Computer Society Annual Symposium on VLSI (ISVLSI'08)*, pages 475 –478, 2008.

[118] M. Guo and L. T. Yang, editors. *New Horizons of Parallel and Distributed Computing*. Springer-Verlag, 2005.

[119] Z. Guz, I. Walter, E. Bolotin, I. Cidon, R. Ginosar, and A. Kolodny. Network Delays and Link Capacities in Application-Specific Wormhole NoCs. *VLSI Design*, 2007.

[120] J. Hamon, L. Fesquet, B. Miscopein, and M. Renaudin. High-Level Time-Accurate Model for the Design of Self-Timed Ring Oscillators. In *Proceedings of the 14th IEEE International Symposium on Asynchronous Circuits and Systems (ASYNC'08)*, pages 29–38. IEEE Computer Society, 2008.

[121] P. K. Harmer, P. D. Williams, G. H. Gunsch, and G. B. Lamont. An Artificial Immune System Architecture for Computer Security Applications. *IEEE Transactions on Evolutionary Computation*, 6(3):252–280, 2002.

[122] M. Hayenga, N. E. Jerger, and M. Lipasti. SCARAB: A Single Cycle Adaptive Routing and Bufferless Network. In *Proceedings of the 42nd Annual International Symposium on Microarchitecture*, pages 244–254, 2009.

[123] S. He, Q. H. Wu, and J. R. Saunders. A Novel Group Search Optimizer Inspired by Animal Behavioural Ecology. In *Proceedings of the IEEE Congress on Evolutionary Computation (CEC'06)*, pages 1272–1278, Vancouver, BC, Canada, July16–21, 2006.

[124] R. Hecht, S. Kubisch, A. Herrholtz, and D. Timmermann. Dynamic Reconfiguration with Hardwired Networks-on-Chip on Future FPGAs. In *Proceedings of the IEEE International Conference on Field Programmable Logic and Applications (FPL'05)*, pages 527–530, Tampere, Finland, August 24–26, 2005.

[125] A. Helmy, L. Pierre, and A. Jantsch. Theorem Proving Techniques for the Formal Verification of NoC Communications with Non-minimal Adaptive Routing. *Design and Diagnostics of Electronic Circuits and Systems*, 0:221–224, 2010.

[126] C. T. Ho and M. T. Raghunath. Concurrency: Practice and Experience. *Efficient Communication Primitives on Hypercubes*, 4(6):427–457, 1992.

[127] C. A. R. Hoare. *Communicating Sequential Processes*. Prentice Hall, 2004.

[128] C. A. R. Hoare, I. J. Hayes, J. He, C. C. Morgan, A. W. Roscoe, J. W. Sanders, I. H. Sorensen, J. M. Spivey, and B. A. Sufrin. Laws of Programming. *Communications of the ACM*, 30(8):672–686, 1987.

[129] M. Hobbs, A. Goscinski, and W. Zhou, editors. *Distributed and Parallel Computing*, volume 3719 of *Lecture Notes in Computer Science*. Springer-Verlag, 2005.

[130] J. H. Holland. *Adaptation in Natural and Artificial Systems*. University of Michigan Press, 1975.

[131] J. S. Hollis and C. Jackson. When Does Network-on-Chip Bypassing Make Sense? In *Proceedings of the 22nd IEEE SoCC Conference*, 2009.

[132] D. A. Holton and J. Sheehan. *The Petersen Graph*. Cambridge University Press, 1993.

[133] J. Hu and R. Marculescu. Energy-Aware Mapping for Tile-Based NoC Architectures Under Performance Constraints. In *Proceedings of the IEEE Asia and South Pacific Design Automation Conference (ASP-DAC'03)*, pages 233–239, January 21–24, 2003.

[134] J. Hu and R. Marculescu. Exploiting the Routing Flexibility for Energy/Performance Aware Mapping of Regular NoC Architectures. In *Proceedings of the DATE'03*, page 10688, 2003.

[135] J. Hu and R. Marculescu. DyAD: Smart Routing for Networks-on-Chip. In *Proceedings of the 41st Conference on Design Automation (DAC'04)*, pages 260–263, 2004.

[136] J. Hu and R. Marculescu. Energy-Aware Communication and Task Scheduling for Network-on-Chip Architectures under Real-Time Constraints. In *Proceedings of the Design Automation Conference (DAC'04)*, page 10234, 2004.

[137] J. Hu and R. Marculescu. Energy and Performance-Aware Mapping for Regular NoC Architectures. *IEEE Transactions on CAD*, 24(4):551–562, 2005.

[138] J. Hu, U. Y. Ogras, and R. Marculescu. System-level Buffer Allocation for Application-specific Networks-on-Chip Router Design. *IEEE Transactions on Computer-Aided Design Of Integrated Circuits And Systems*, 25:2919–2933, 2006.

[139] M. Hubner, M. Ullmann, L. Braun, A. Klausmann, and J. Becker. Scalable Application-Dependent Network on Chip Adaptivity for Dynamical Reconfigurable Real-Time Systems. In *Proceedings of International Conference on Field Programmable Logic and Applications (FPL'04)*, pages 1037–1041, 2004.

[140] A. Ivanov and G. De Micheli. Guest Editors' Introduction: The Network-on-Chip Paradigm in Practice and Research. In *Proceedings of the IEEE Design & Test of Computers*, pages 399–403. IEEE, 2000.

[141] A. Jantsch. *Modeling Embedded Systems and SoC's: Concurrency and Time in Models of Computation*. Morgan Kaufmann, 2003.

[142] A. Jantsch and H. Tenhunen, editors. *Networks on Chip*. Kluwer Academic Publishers, February 2003.

[143] J. Jaros. *Evolutionary Design of Collective Communications on Wormhole Networks*. PhD thesis, Faculty of Information Technology, Brno University of Technology, Brno, Czech Rep., 2010.

[144] N. Enright Jerger, L.-S. Peh, and M. H. Lipasti. Circuit-Switched Coherence. In *Proceedings of the 2nd Annual Network on Chip Symposium*, 2008.

[145] N. K. Jerne. Towards a Network Theory of the Immune System. *Ann. Immunol. (Inst. Pasteur) 125C*, pages 373–389, 1974.

[146] A. Kahng, B. Li, L. S. Peh, and K. Samadi. ORION 2.0: A Fast and Accurate NoC Power and Area Model for Early-Stage Design Space Exploration. In *Proceedings of the IEEE Design, Automation and Test in Europe Conference and Exhibition (DATE'09)*, pages 423–428, April 20–24, 2009.

[147] M. Kamali, L. Laibinis, L. Petre, and K. Sere. Self-Recovering Sensor-Actor Networks. In *Proceedings of the 9th International Workshop on the Foundations of Coordination Languages and Software Architectures (FOCLASA'10)*, Electronic Proceedings in Theoretical Computer Science (EPTCS), 2010.

[148] M. Kandemir, O. Ozturk, and S. P. Muralidhara. Dynamic Thread and Data Mapping for NoC-Based CMPs. In *Proceedings of the 46th Annual Design Automation Conference (DAC'09)*, pages 852–857, 2009.

[149] F. Karim and A. Nguyen. An Interconnect Architecture for Networking Systems on Chips. In *Proceedings of the IEEE Micro*, pages 36–45. IEEE, 2002.

[150] S. Katz. A Superimposition Control Construct for Distributed Systems. *ACM Transactions on Programming Languages and Systems*, 15(2):337–356, 1993.

[151] J. Kennedy and R. C. Eberhart. Particle Swarm Optimization. In *Proceedings of the IEEE International Conference on Neural Networks (ICNN'95)*, pages 12–13, Perth, Australia, December 1995.

[152] J. O. Kephart and D. M. Chess. The Vision of Autonomic Computing. *Computer*, 36(1):41–50, 2003.

[153] K. Keutzer, A. R. Newton, J. M. Rabaey, and A. Sangiovanni-Vincentelli. System-Level Design: Orthogonalization of Concerns and Platform-Based Design. *IEEE Transactions on CAD*, 19(12):1523–1543, December 2000.

[154] D. Kim, K. Lee, S. Lee, and H. Yoo. A Reconfigurable Crossbar Switch with Adaptive Bandwidth Control for Networks-on-Chip. In *Proceedings of the IEEE International Symposium on Circuits and Systems (ISCAS'05)*, volume 3, pages 2369–2372, 2005.

[155] J. Kim, M. Lee, W. Kim, J. Chang, Y. Bae, and H. Cho. Performance Analysis of NoC Structure Based on Star-Mesh Topology. In *Proceedings of the International SoC Design Conference (ISOCC'08)*, volume 2, pages 162–165, Busan, Korea, November 24–25, 2008.

[156] J. Kim, D. Park, T. Theocharides, N. Vijaykrishnan, and Chita R. Das. A Low Latency Router Supporting Adaptivity for On-chip Interconnects. In *Proceedings of the 42nd annual Design Automation Conference (DAC'05)*, pages 559–564, 2005.

[157] W. Kim, M. S. Gupta, G. Y. Wei, and D. Brooks. System Level Analysis of Fast, Per-Core DVFS Using on-Chip Switching Regulators. In *Proceedings of HPCA'08*, pages 123–134, February 16–20 2008.

[158] A. K. Kodi, A. Sarathy, and A. Louri. Adaptive Channel Buffers in On-Chip Interconnection Networks–A Power and Performance Analysis. *IEEE Transactions on Computers*, 57(9):1169–1181, 2008.

[159] A. K. Kodi, A. Sarathy, and A. Louri. iDEAL: Inter-Router Dual-Function Energy and Area-Efficient Links for Network-on-Chip (NoC) Architectures. In *Proceedings of ISCA'08*, pages 241–250, 2008.

[160] J. R. Koza. *Genetic Programming: On the Programming of Computers by Means of Natural Selection*. MIT Press, December 1992.

[161] N. Koziris, M. Romesis, P. Tsanakas, and G. Papakonstantinou. An Efficient Algorithm for the Physical Mapping of Clustered Taskgraphs onto Multiprocessor Architectures. In *Proceedings of the 8th IEEE Euromicro Workshop on Parallel and Distributed Processing (EURO-PDP'00)*, pages 406–413, Rhodos, Greece, January 19–21, 2000.

[162] M. Krstic et al. System Integration by Request-Driven GALS Design. In *Proceedings of the IEE Computers & Digital Techniques*, volume 153, pages 362–372, September 2006.

[163] V. Kumar, A. Grama, A. Gupta, and G. Karypis. *Introduction to Parallel Computing: Design and Analysis of Algorithms*. Benjamin/Cummings Press, 1994.

[164] A. Kumary, L-S. Pehy, P. Kunduz, and N. K. Jhay. Express Virtual Channels: Towards the Ideal Interconnection Fabric. In *Proceedings of the 34th Annual International Symposium on Computer Architecture*, pages 150–161, 2007.

[165] W. Kuzmicz, E. Piwowarska, A. Pfitzner, and D. Kasprowicz. Static Power Consumption in Nano-CMOS Circuits: Physics and Modelling. In *Proceedings of the 14th International Conference Mixed Design of Integrated Circuits and Systems*, pages 163–168, June 2007.

[166] M. Lai, Z. Wang, L. Gao, H. Lu, and K. Dai. A Dynamically Allocated Virtual Channel Architecture with Congestion Awareness for On-Chip Routers. In *Proceedings of the DAC'08*, pages 630–633, 2008.

[167] Y. Lan, S. Lo, Y. Lin, Y. Hu, and S. Chen. BiNoC: A Bidirectional NoC Architecture with Dynamic Self-Reconfigurable Channel. In *Proceedings of the 3rd ACM/IEEE International Symposium on Networks-on-Chip (NoCS'09)*, pages 266 –275, 2009.

[168] P. Larraaga and J.A. Loazano. *Estimation of Distribut ion Algorithms, A New Tool for Evolutionary Computation*. Kluwer Academic Publishers, 2002.

[169] K. Latif, M. Niazi, H. Tenhunen, T. Seceleanu, and S. Sezer. Application Development Flow for on-Chip Distributed Architectures. In *Proceedings of the IEEE International SOC Conference*, pages 163 –168, 2008.

[170] D. A. Lawrie. Access and Alignment of Data in an Array Processor. *IEEE Transactions on Computers*, 24:1145–1155, 1975.

[171] F. W. Lawvere and S. H. Schanuel. *Conceptual Mathematics: A First Introduction to Categories*. Cambridge University Press, 1997.

[172] G. Leary, K. Srinivasan, K. Mehta, and K. Chatha. Design of Network-on-Chip Architectures With a Genetic Algorithm-Based Technique. *IEEE Transactions on Very Large Scale Integration (VLSI) Systems*, 17(5):674–687, May 2009.

[173] E. A. Lee. Cyber-Physical Systems – Are Computing Foundations Adequate? In *NSF Workshop On Cyber-Physical Systems: Research Motivation, Techniques and Roadmap*, 2006.

[174] E. A. Lee. Cyber-Physical Systems: Design Challenges. Technical Report UCB/EECS-2008-8, EECS Department, University of California, Berkeley, January 2008.

[175] R. Lee, editor. *Software Engineering, Artificial Intelligence, Networking and Parallel/Distributed Computing*, volume 149 of *Studies in Computational Intelligence*. Springer-Verlag, 2008.

[176] T. Lehtonen, P. Liljeberg, and J. Plosila. Online Reconfigurable Self-Timed Links for Fault Tolerant NoC. *VLSI Design*, 2007:13, 2007.

[177] T. Lei and S. Kumar. A Two-Step Genetic Algorithm for Mapping Task Graphs to a Network on Chip Architecture. In *Proceedings of the Euromicro Symposium on Digital Systems Design*, pages 180–189, 2003.

[178] C. E. Leiserson. Fat-Trees: Universal Networks for Hardware Efficient Supercomputing. *IEEE Transactions on Computers*, 34:892–901, 1985.

[179] P. De Lellis and M. Di Bernardo. Robustness of Local Adaptive Synchronization Strategies to Topological Variations and Delays. In *Proceedings of the International Symposium on Circuit and Systems*, pages 1609–1612, 2009.

[180] M. Leuschel and M. Butler. ProB: A Model Checker for B. In *Proceedings of FME'03*, volume 2805 of *Lecture Notes in Computer Science*, pages 855–874. Springer-Verlag, 2003.

[181] F. W. Levi. *Finite Geometrical Systems*. University of Calcutta, 1942.

[182] M. Levine. Categorical Algebra. In G. Benkart, T. S. Ratiu, H. A. Masur, and M. Renardy, editors, *Mixed Motives*, volume 57 of *Mathematical Surveys and Monographs*, chapter I, II, II of Part II, pages 373–499. American Mathematical Society, 1998.

[183] J. Li and U. Aickelin. A Bayesian Optimization Algorithm for the Nurse Scheduling Problem. In *Proceedings of the Congress on Evolutionary Computation (CEC'03)*, pages 2149–2156. IEEE Computer Society, 2003.

[184] M. Li, Q. A. Zeng, and W. B. Jone. DyXY-A Proximity Congestion-Aware Deadlock-Free Dynamic Routing Method for Network on Chip. In *Proceedings of the DAC'06*, pages 849–859, 2006.

[185] K. M. Liew, H. Shen, S. See, W. Cai, P. Fan, and S. Horiguchi, editors. *Parallel and Distributed Computing: Applications and Technologies*, volume 3320 of *Lecture Notes in Computer Science*. Springer-Verlag, 2005.

[186] G. M. Link and N. Vijaykrishnan. Hotspot Prevention through Runtime Re-configuration in Network-on-Chip. In *Proceedings of the Conference on De-sign, Automation and Test in Europe*, volume 1, pages 648–649, 2005.

[187] B. G. Liptak. *Instrument Engineers' Handbook: Process Control and Opti-mization*. CRC Press, September 2005.

[188] P. Lotfi-Kamran, A. M. Rahmani, M. Daneshtalab, A. Afzali-Kusha, and Z. Navabi. EDXY – A Low Cost Congestion-Aware Routing Algorithm for Network-on-Chips. *Journal of Syst. Archit.*, 56(7):256–264, 2010.

[189] Y. H. Lu and G. De Micheli. Comparing System-Level Power Management Policies. *IEEE Design and Test of Computers*, 18:10–19, 2001.

[190] Z. Lu, L. Xia, and A. Jantsch. Cluster-Based Simulated Annealing for Map-ping Cores onto 2D Mesh Networks on Chip. In *Proceedings of the 11th IEEE International Workshop on Design and Diagnostics of Electronics Cir-cuits and Systems (DDECS'08)*, pages 1–6, Bratislava, Slovakia, April 16–18, 2008.

[191] G. Magklis, G. Semeraro, D. H. Albonesi, S. G. Dropsho, S. Dwarkadas, and M. L. Scott. Dynamic Frequency and Voltage Scaling for a Multiple-Clock-Domain Microprocessor. *IEEE Micro*, 23:62–68, 2003.

[192] D. Marculescu and E. Talpes. Energy Awareness and Uncertainty in Microarchitecture-Level Design. *IEEE Micro*, 25(5):64–76, 2005.

[193] R. Marculescu. Networks-on-Chip: The Quest for on-Chip Fault-Tolerant Communication. In *Proceedings of the IEEE Computer Society Annual Sym-posium on VLSI (ISVLSI'03)*. IEEE, 2003.

[194] R. Marculescu and P. Bogdan. The Chip Is the Network: Toward a Science of Network-on-Chip Design. *Foundations and Trends in Electronic Design Automation*, 2(4):371–461, 2007.

[195] R. Marculescu, U. Y. Ogras, Li-Shiuan Peh, N. E. Jerger, and Y. Hoskote. Outstanding Research Problems in NoC Design: System, Microarchitecture, and Circuit Perspectives. *IEEE Transactions on Computer-Aided Design of Integrated Circuits and Systems*, 28(1):3–21, January 2009.

[196] B. Marius and B. Howard. An Integrated Specification Logic for Cyber-Physical Systems. In *Proceedings of the 14th IEEE International Conference on Engineering of Complex Computer Systems*, pages 291–300, 2009.

[197] MATLAB® R2007a, Genetic Algorithm and Direct Search Toolbox 2.1. The MathWorks Inc. http://www.mathworks.com/, Natick, MA, 2007.

[198] P. K. McKinley and C. Trefftz. Efficient Broadcast in All-Port Wormhole-Routed Hypercubes. In *Proceedings of International Conference on Parallel Processing*, volume 11, pages 288–291, 2003.

[199] Z. Michalewiz. *Genetic Algorithm + Data Structure = Evolution Programs.* Springer-Verlag, March 1996.

[200] R. Michel. A Categorical Approach to Distributed Systems Expressibility and Knowledge. In *Proceedings of the 8th annual ACM Symposium on Principles of distributed computing (PODC'89),* pages 129–143. ACM, 1989.

[201] G. De Micheli and L. Benini. Networks on Chip: A New Paradigm for Systems on Chip Design. In *DATE '02: Proceedings of the conference on Design, automation and test in Europe,* page 418. IEEE Computer Society, 2002.

[202] G. Michelogiannakis, J. Balfour, and W. J. Dally. Elastic-Buffer Flow Control for On-Chip Networks. In *Proceedings of the International Symposium on High-Performance Computer Architecture.,* pages 151–162, 2009.

[203] R. Milner. *Communication and Concurrency.* International Series in Computer Science. Prentice Hall, December 1989.

[204] N. Minar, M. Gray, O. Roup, R. Krikorian, and P. Maes. Hive: Distributed Agents for Networking Things. *IEEE Concurrency,* 8(2):24–33, 2000.

[205] A. K. Mishra, R. Das, S. Eachempati, R. Iyer, N. Vijaykrishnan, and C. R. Das. A Case for Dynamic Frequency Tuning in On-Chip Networks. In *Proceedings of the 42nd Annual IEEE/ACM International Symposium on Microarchitecture (MICRO'09),* pages 292–303, 2009.

[206] M. Modarressi and M. Arjomand H. Sarbazi-Azad. A Hybrid Packet-Circuit Switched On-Chip Network Based on SDM. In *Proceedings of DATE'09,* pages 566–569, 2009.

[207] L. Moller, I. Grehs, E. Carvalho, R. Soares, N. Calazans, and F. Moraes. A NoC-based Infrastructure to Enable Dynamic Self Reconfigurable Systems. In *Proceedings of the 3rd International Workshop on Reconfigurable Communication-Centric Systems-on-Chip (ReCoSoC'07),* pages 23–30, Montpellier, France, June 18–20, 2007.

[208] M. Moore and T. Suda. A Decentralized and Self-organizing Discovery Mechanism. In *Proceedings of the 1st Annual Symposium on Autonomous Intelligent Networks and Systems,* 2002.

[209] A. A. Morgan, H. Elmiligi, M. W. El-Kharashi, and F. Gebali. Application-Specific Networks-on-Chip Topology Customization Using Network Partitioning. In *Proceedings of the 1st International Forum on Next-Generation Multicore/Manycore Technologies (IFMT'08),* pages 1–6, Cairo, Egypt, November 24–25, 2008.

[210] A. A. Morgan, H. Elmiligi, M. W. El-Kharashi, and F. Gebali. Networks-on-Chip Topology Generation Techniques: Area and Delay Evaluation. In *Proceedings of the 3rd IEEE International Design and Test Workshop (IDT'08),* pages 33–38, Monastir, Tunisia, December 20–22, 2008.

[211] A. A. Morgan, H. Elmiligi, M. W. El-Kharashi, and F. Gebali. Area-Aware Topology Generation for Application-Specific Networks-on-Chip Using Network Partitioning. In *Proceedings of the 2009 IEEE Pacific Rim Conference on Communications, Computers and Signal Processing (PACRIM'09)*, pages 979–984, Victoria, BC, Canada, August 23–26, 2009.

[212] A. A. Morgan, H. Elmiligi, M. W. El-Kharashi, and F. Gebali. Multi-Objective Optimization for Networks-on-Chip Architectures Using Genetic Algorithms. In *Proceedings of the IEEE International Symposium on Circuits and Systems (ISCAS'10)*, pages 3725–3728, Paris, France, May 30–June 2, 2010.

[213] M. Morvarid, M. Fathy, R. Berangi, and A. Khademzadeh. IIIModes: New Efficient Dynamic Routing Algorithm for Network on Chips. In *Proceedings of the 4th International Multi-Conference on Computing in the Global Information Technology*, pages 57–62, 2009.

[214] T. Moscibroda and O. Mutlu. A Case for Bufferless Routing in On-Chip Networks. In *Proceedings of the 36th International Symposium on Computer Architecture (ISCA'09)*, pages 196–207, 2009.

[215] H. Muhlenbein and G. Paas. From Recombination of Genes to the Estimation of Distributions I. Binary Parameters. In *Parallel Problem Solving from Nature (PPSN)*, volume 1411 of *Lecture Notes in Computer Science*, pages 178–187, 1996.

[216] R. Mullins and S. Moore. Demystifying Data-Driven and Pausible Clocking Schemes. In *Proceedings of the 13th IEEE International Symposium on Asynchronous Circuits and Systems (ASYNC'07)*, pages 175–185. IEEE Computer Society, 2007.

[217] R. Mullins, A. West, and S. Moore. Low-Latency Virtual-Channel Routers for on-Chip Networks. In *Proceedings of the 31st Annual International Symposium on Computer Architecture (ISCA'04)*, pages 188–197, 2004.

[218] R. Munafo. *The Diameter-Degree Problem*, 2008.

[219] S. Murali, L. Benini, and G. Micheli. Mapping and Physical Planning of Networks-on Chip Architectures with Quality-of-Service Guarantees. In *Proceedings of the 2005 Conference on Asia South Pacific Design Automation*, pages 27–32, 2005.

[220] S. Murali and G. De Micheli. Bandwidth-Constrained Mapping of Cores onto NoC Architectures. In *Proceedings of the IEEE Design, Automation and Test in Europe Conference and Exhibition (DATE'04)*, volume 2, pages 896–901, Paris, France, February 16–20, 2004.

[221] S. Murali and G. De Micheli. SUNMAP: A Tool for Automatic Topology Selection and Generation for NoCs. In *Proceedings of the 41st IEEE/ACM Design Automation Conference (DAC'04)*, pages 914–919, San Diego, CA, June 7–11, 2004.

[222] S. Murali, C. Seiculescu, L. Benini, and G. De Micheli. Synthesis of Networks on Chips for 3D Systems on Chips. In *Proceedings of the 2009 Conference on Asia and South Pacific Design Automation*, pages 242–247. IEEE, 2009.

[223] K. Murty. *Linear and Combinatorial Programming*. John Wiley, 1976.

[224] J. Muttersbach, T. Villiger, and W. Fichtner. Practical Design of Globally-Asynchronous Locally-Synchronous Systems. In *Proceedings of the 6th International Symposium on Advanced Research in Asynchronous Circuits and Systems (ASYNC'00)*, page 52. IEEE Computer Society, 2000.

[225] M. H. Neishaburi and Z. Zilic. Reliability Aware NoC Router Architecture Using Input Channel Buffer Sharing. In *Proceedings of the ACM Great Lakes Symposium on VLSI*, pages 511–516, 2009.

[226] L. M. Ni and P. K. McKinley. A Survey of Wormhole Routing Techniques in Direct Networks. *Computer*, 26:62–76, 1993.

[227] M. Nichschas and U. Brinkschulte. Decentralized Task Allocation in an Organic Real-Time Middleware: An Auction-Based Approach. In *Proceedings of the 2009 Symposia and Workshops on Ubiquitous, Autonomic and Trusted Computing*, pages 574–579. IEEE Computer Society, 2009.

[228] A. Nicoli. *Achieving Yield in the Nanometer Age*. Technical Report, Mentor Graphics Corp., December 2007.

[229] C. A. Nicopoulos, D. Park, J. Kim, N. Vijaykrishnan, M. S. Yousif, and C. R. Das. ViChaR: A Dynamic Virtual Channel Regulator for Network-on-Chip Routers. In *Proceedings of the 39th Annual IEEE/ACM International Symposium on Microarchitecture (MICRO'06)*, pages 333–346, 2006.

[230] N. Nupairoj and L. M. Ni. *Benchmarking of Multicast Communication Services*. Technical Report MSU-CPS-ACS-103, Michigan State University, 1995.

[231] J. Nurmi, H. Tenhunen, J. Isoaho, and A. Jantsch, editors. *Interconnect-Centric Design for Advanced SoC and NoC*. Kluwer Academic Publishers, 2004.

[232] J. Ocenasek. *Parallel Estimation of Distribution Algorithms*. PhD thesis, Faculty of Information Technology, Brno University of Technology, Brno, Czech Rep., 2002.

[233] U. Y. Ogras, J. Hu, and R. Marculescu. Key Research Problems in NoC Design: A Holistic Perspective. In *Proceedings of the Int'l. Conf. on Hardware/Software Codesign and System Synthesis*, September 2005.

[234] U.Y. Ogras and R. Marculescu. It's a Small World after All: NoC Performance Optimization via Long-Range Link Insertion. *IEEE Transactions on Very Large Scale Integration (VLSI) Systems*, 14(7):693–706, July 2006.

[235] V. G. Oklobdzija. Clocking and Clocked Storage Elements in a Multi-Gigahertz Environment. *IBM Journal of Res. Dev.*, 47(5–6):567–583, 2003.

[236] L. Ost, A. Mello, J. Palma, F. G. Moraes, and N. Calazans. MAIA: A Framework for Networks on Chip Generation and Verification. In *Proceedings of the ASP-DAC Conference*, pages 49–52, 2005.

[237] P. P. Pande, C. Grecu, M. Jones, A. Ivanov, and R. Saleh. Performance Evaluation and Design Tradeoffs for Network-on-Chip Interconnect Architectures. *IEEE Transactions on Computers*, 54(8):1025–1040, August 2005.

[238] B. Pangrle and S. Kapoor. *Leakage Power at 90nm and Below*. Technical Report, Synopsis Inc., May 2005.

[239] V. Pavlidis and E. Friedman. 3-D Topologies for Networks-on-Chip. *IEEE Transactions on Very Large Scale Integration (VLSI) Systems*, 15(10):1081–1090, October 2007.

[240] M. Pedram. Design Technologies for Low Power VLSI. In *Encyclopedia of Computer Science and Technology*, pages 73–96. Marcel Dekker, Inc., 1997.

[241] L. S. Peh and W. J. Dally. A Delay Model for Router Microarchitecture. *IEEE Micro*, 21:26–34, 2001.

[242] L. Petre, K. Sere, and M. Waldén. A Language for Modeling Network Availability. In *Proceedings of the 8th International Conference on Formal Engineering Methods (ICFEM'2006)*, volume 4260 of *Lecture Notes in Computer Science*, pages 639–659. Springer-Verlag, 2006.

[243] L. Petre, K. Sere, and M. Waldén. *Ensuring Correctness of Network Services with MIDAS*. Technical Report 938, Turku Centre for Computer Science (TUCS), May 2009.

[244] L. Pierre, D. Borrione, A. Helmy, and J. Schmaltz. A Generic Model for Formally Verifying NoC Communication Architectures: A Case Study. In *Proceedings of the ACM/IEEE Int'l. Symp. on Networks-on-Chip (NOCS'07)*, May 2007.

[245] C. Piguet, editor. *Low-Power Electronics Design*. CRC Press, 2005.

[246] A. Pinto, L. P. Carloni, and A. L. Sangiovanni-Vincentelli. Efficient Synthesis of Networks on Chip. In *Proceedings of the 21st International Conference on Computer Design (ICCD'03)*, page 146, 2003.

[247] T. Pionteck, C. Albrecht, and R. Koch. A Dynamically Reconfigurable Packet-Switched Network-on-Chip. In *Proceedings of the IEEE Design, Automation and Test in Europe Conference and Exhibition (DATE'06)*, volume 1, pages 136–137, Munich, Germany, March 6–10, 2006.

[248] T. Pionteck, R. Koch, and C. Albrecht. Applying Partial Reconfiguration to Networks-on-Chips. In *Proceedings of FPL'06*, pages 1–6, 2006.

[249] J. Pouwelse, K. Langendoen, and H. Sips. Dynamic Voltage Scaling on a Low-Power Microprocessor. In *Proceedings of the 7th Annual International Conference on Mobile Computing and Networking (MobiCom'01)*, pages 251–259. ACM, 2001.

[250] Predictive Technology Model (PTM). Arizona State University. Available: http://ptm.asu.edu/, Last accessed August 2010.

[251] B. R. Quinton and S. J. E. Wilton. Programmable Logic Core Enhancements for High-Speed on-Chip Interfaces. *IEEE Transactions on Very Large Scale Integration Systems*, 17(9):1334–1339, 2009.

[252] J. M. Rabaey. Scaling the Power Wall: Revisiting the Low-Power Design Rules. Keynote speech at SoC'07 Symposium, November 2007.

[253] J. M. Rabaey. System-on-Chip-Challenges in the Deep-Sub-Micron Era: A Case for the Network-on-a-Chip. *In J. Nurmi et al. (Ed.), Interconnect-Centric Design for Advanced SoC and NoC*, pages 3–24, Springer, 2005.

[254] A. M. Rahmani, A. Afzali-Kusha, and M. Pedram. NED: A Novel Synthetic Traffic Pattern for Power/Performance analysis of Network-on-chips Using Negative Exponential Distribution. *Journal of Low Power Electronics*, 5:396–405, 2009.

[255] A. M. Rahmani, M. Daneshtalab, A. Afzali-Kousha, and M. Pedram. Forecasting-Based Dynamic Virtual Channels Allocation for Power Optimization of Network-on-Chips. In *Proceedings of the 22nd Int'l Conference on VLSI Design*, pages 151–156, 2009.

[256] A. M. Rahmani, M. Daneshtalab, A. Afzali-Kusha, and M. Pedram. Forecasting-Based Dynamic Virtual Channel Management for Power Reduction in Network-on-Chips. *Journal of Low Power Electronics*, 5:385–395, 2009.

[257] P. Rantala, J. Isoaho, and H. Tenhunen. Agent-Based Reconfigurability for Fault-Tolerance in Network-on-Chip. In *Proceedings of the International Conference on Engineering of Reconfigurable Systems and Algorithms*, pages 207–210, 2007.

[258] E. Rijpkema, K. Goossens, A. Radulescu, J. Dielissen, J. Van Meerbergen, P. Wielage, and E. Waterlander. Tradeoffs in the Design of a Router with both Guaranteed and Best-Effort Services for Networks on Chip. In *Proceedings of Computers and Digital Techniques*, volume 150, pages 294–302, 2003.

[259] D. Rios-Arambula, A. Buhrig, G. Sicard, and M. Renaudin. On the Use of Feedback Systems to Dynamically Control the Supply Voltage of Low-Power Circuits. *Journal of Low Power Electronics*, 2(1):45–55, 2006.

[260] J. Rius, M. Meijer, and J. P. de Gyvez. An Activity Monitor for Power/Performance Tuning of CMOS Digital Circuits. *Journal of Low Power Electronics*, 2(1):80–86, 2006.

[261] B. F. Romanescu, M. E. Bauer, D. J. Sorin, and S. Ozev. Reducing the Impact of Process Variability with Prefetching and Criticality-Based Resource Allocation. In *Proceedings of the 16th International Conference on Parallel Architecture and Compilation Techniques (PACT'07)*, page 424. IEEE Computer Society, 2007.

[262] S. J. Russell and P. Norvig. *Artificial Intelligence: A Modern Approach*. Prentice Hall, 1995.

[263] Y. Saad and M. H. Schultz. Topological Properties of Hypercubes. *IEEE Transactions on Computers*, 37:867–872, 1988.

[264] G. Al Sammane, J. Schmaltz, and D. Borrione. Formal Verification of On-Chip Networking. In *Proceedings of the 1st International Conference on Information & Communication Technologies: from Theory to Applications (ICTTA'04)*, 2004.

[265] A. Sangiovanni-Vincentelli and G. Martin. Platform-Based Design and Software Design Methodology for Embedded Systems. *IEEE Design Test*, 18(6):23–33, 2001.

[266] S. V. Sathish, E. F. Graham, and D. Jack. Automatically Tuned Collective Communications. In *Proceedings of the ACM/IEEE Conference on Supercomputing*, page 3, 2000.

[267] T. Schattkowsky and W. Muller. Model-Based Design of Embedded Systems. In *Proceedings of the 7th IEEE International Symposium on Object-Oriented Real-Time Distributed Computing*, pages 113–128, 2004.

[268] J. Schmaltz and D. Borrione. A Functional Approach to the Formal Specification of Networks on Chip. In *Proceedings of Formal Methods in Computer-Aided Design (FMCAD'04)*, 2004.

[269] J. Schmaltz and D. Borrione. A Functional Specification and Validation Model for Networks on Chip in the ACL2 Logic. In *Proceedings of the 5th International Workshop on the ACL2 Theorem Prover and its Applications (ACL2'04)*, 2004.

[270] L. Sekanina. Nanostructures and Bio-Inspired Computer Enegineering. In *Proceedings of Nano'02*, pages 233–236, 2002.

[271] G. Semeraro, G. Magklis, R. Balasubramonian, D. H. Albonesi, S. Dwarkadas, and M. L. Scott. Energy-Efficient Processor Design Using Multiple Clock Domains with Dynamic Voltage and Frequency Scaling. In *Proceedings of the 8th International Symposium on High-Performance Computer Architecture*, pages 29–40. IEEE Computer Society, February 2002.

[272] D. Serpanos and J. Henkel. Dependability and Security Will Change Embedded Computing. *Computer*, 41(1):103–105, 2008.

[273] M. Sgroi, M. Sheets, A. Mihal, K. Keutzer, S. Malik, J. Rabaey, and A. Sangiovanni-Vincentelli. Addressing the System-on-a-Chip Interconnect Goes through Communication-based Design. In *Proceedings of Design Automation Conference*, pages 667–672, June 2001.

[274] W. Shen, C. Chao, Y. Lien, and A. Wu. A New Binomial Mapping and Optimization Algorithm for Reduced-Complexity Mesh-Based On-Chip Network. In *Proceedings of the IEEE International Symposium on Networks-on-Chip (NOCS'07)*, pages 317–322, May 7–9, 2007.

[275] K. G. Shin and S. W. Daniel. Analysis and Implementation of Hybrid Switching. *IEEE Transactions on Computers*, 45(6):684–692, 1996.

[276] T. Simunic, L. Benini, and G. De Micheli. Energy-Efficient Design of Battery-Powered Embedded Systems. *IEEE Transactions on VLSI*, 9(1):15–28, February 2001.

[277] K. Skadron, T. Abdelzaher, and M. R. Stan. Control-Theoretic Techniques and Thermal-RC Modeling for Accurate and Localized Dynamic Thermal Management. In *Proceedings of the 8th International Symposium on High-Performance Computer Architecture (HPCA'02)*, pages 17–28. IEEE Computer Society, 2002.

[278] L. T. Smit, G. J. M. Smit, J. L. Hurink, H. Broersma, D. Paulusma, and P. T. Wolkotte. Run-Time Mapping of Applications to a Heterogeneous Reconfigurable Tiled System on Chip Architecture. In *Proceedings of the IEEE International Conference on Field-Programmable Technology (FPT'04)*, pages 421–424, Brisbane, Australia, 6–8 December 2004.

[279] A. Somayaji, S. Hofmeyr, and S. Forrest. Principles of a Computer Immune System. In *Proceedings of the 2nd New Security Paradigms Workshop*, pages 75–82, 1997.

[280] V. Soteriou and L.-S. Peh. Exploring the Design Space of Self-regulating Power-aware On/Off Interconnection Networks. *IEEE Transactions on Parallel and Distributed Systems*, 18(3):393–408, 2007.

[281] V. Soteriou, R. S. Ramanujam, B. Lin, and L. Peh. A High-Throughput Distributed Shared-Buffer NoC Router. *Computer Architecture Letters*, 8(1):21–24, January 2009.

[282] K. Srinivasan and K. Chatha. ISIS: A Genetic Algorithm Based Technique for Custom on-Chip Interconnection Network Synthesis. In *Proceedings of the 18th IEEE International Conference on VLSI Design (VLSID'05)*, pages 623–628, Kolkata, India, January 3–7, 2005.

[283] K. Srinivasan and K. Chatha. A Low Complexity Heuristic for Design of Custom Network-on-Chip Architectures. In *Proceedings of the IEEE Design, Automation and Test in Europe Conference (DATE'06)*, volume 1, pages 1–6, Munich, Germany, March 6–10, 2006.

[284] K. Srinivasan, K. Chatha, and G. Konjevod. Linear-Programming-Based Techniques for Synthesis of Network-on-Chip Architectures. *IEEE Transactions on Very Large Scale Integration (VLSI) Systems*, 14(4):407–420, April 2006.

[285] M. B. Stensgaard and J. Sparso. ReNoC: A Network-on-Chip Architecture with Reconfigurable Topology. In *Proceedings of the 2nd ACM/IEEE International Symposium on Networks-on-Chip*, pages 55–64, 2008.

[286] L. C. Stewart and D. Gingold. *A New Generation of Cluster Interconnect*. White Paper, SiCortex Inc., 2006.

[287] M. Stork. Digital Building Block for Frequency Synthesizer and Fractional Phase Locked Loops. In *Proceedings of the IEEE Mobile Future and Symposium on Trends in Communications*, pages 126–129, October 2003.

[288] T. Streichert, M. Glass, R. Wanka, C. Haubelt, and J. Teich. Topology-Aware Replica Placement in Fault-Tolerant Embedded Networks. In *Proceedings of the Architecture of Computing Systems (ARCS'08)*, volume 4934 of *Lecture Notes in Computer Science*, pages 23–37. Springer-Verlag, 2008.

[289] S. Suboh. *Towards an Adaptive Interconnect for System-on-Chip*. Thesis Report, HPC Laboratory, The George Washington University, USA, 2010.

[290] S. Suboh, M. Bakhouya, and T. El-Ghazawi. Simulation and Evaluation of On-Chip Interconnect Architectures: 2D Mesh, Spidergon, and WK-Recursive Networks. In *Proceedings of NoCS'08*, pages 205–206, 2008.

[291] S. Suboh, M. Bakhouya, J. Gaber, and T. El-Ghazawi. An Interconnection Architecture for Network-on-Chip Systems. *Telecom. Systems*, 37(1-3):137–144, 2008.

[292] D. Sylvester, D. Blaauw, and E. Karl. ElastIC: An Adaptive Self-Healing Architecture for Unpredictable Silicon. *IEEE Design & Test of Computers*, 23(6):484–490, 2006.

[293] W. J. Tang and Q. H. Wu. Biologically Inspired Optimization: A Review. *Transactions of the Institute of Measurement and Control*, 31(6):495–515, 2009.

[294] H. Tenhunen. Autonomous NoC (aNOC): A Co-operative Project between KTH and University of Turku. [Online]. http://www.mpsocforum.org/2006/slides/Tenhunen.pdf, Last accessed August 2010.

[295] C. Teuscher. *On the State-of-the-Art of POETIC Machines*. Technical Report 01/375, EPFL, Computer Science Department, 2001.

[296] D. N. Truong, W. H. Cheng, T. Mohsenin, Z. Yu, A. T. Jacobson, G. Landge, et al. A 167-Processor Computational Platform in 65 nm CMOS. *IEEE Journal of Solid State Circuits*, 44(4):1130–1144, 2009.

[297] Y. Tsai and P. K. McKinley. An Extended Dominating Node Approach to Broadcast and Global Combine in Multiport Wormhole-Routed Mesh Networks. *IEEE Transactions on Parallel and Distributed Systems*, 8(1):427–457, 1997.

[298] J. Tschanz, J. Kao, S. Narendra, et al. Adaptive Body Bias for Reducing Impacts of Die-to-Die and within-Die Parameter Variations on Microprocessor Frequency and Leakage. *IEEE Journal of Solid-state Circuits*, 37(11):1396–1402, November 2002.

[299] L. Tsiopoulos, L. Petre, and K. Sere. Model-Based Placement of Pipelines Applications on a NoC, Submitted 2010.

[300] P. Tvrdik. Parallel Systems and Algorithms. *Lecture Notes in Computer Science*, page 177, 1997.

[301] N. F. Tzeng. Reliable Butterfly Distributed-Memory Multiprocessors. *IEEE Transactions on computers*, 43(9):1004–1013, 1994.

[302] Indiana University. Open MPI: Open Source High Performance Computing, 2009.

[303] A. Upegui, Y. Thoma, E. Sanchez, A. Perez-Uribe, J. M. Moreno, and J. Madrenas. The Perplexus Bio-Inspired Reconfigurable Circuit. In *Proceedings of the 2nd NASA/ESA Conference on Adaptive Hardware and Systems (AHS'07)*, pages 600–605, 2007.

[304] T. Valtonen, T. Nurmi, J. Isoaho, and H. Tenhunen. An Autonomous Error-Tolerant Cell for Scalable Network on Chip Architectures. In *Proceedings of the 19th Nordic Microelectronics Conference (NORCHIP'01)*, pages 198–203, Kista, Sweden, November 12-13, 2001.

[305] S.R. Vangal, J. Howard, G. Ruhl, S. Dighe, H. Wilson, J. Tschanz, et al. An 80-Tile Sub-100-W TeraFLOPS Processor in 65-nm CMOS. *IEEE JSSC*, 43(1):29–41, 2008.

[306] F. J. Varela and A. Coutinho. Second Generation Immune Networks. *Immunology Today*, 12(5):159–166, 1991.

[307] A. Varma, B. Ganesh, M. Sen, S. R. Choudhury, L. Srinivasan, and B. Jacob. A Control-Theoretic Approach to Dynamic Voltage Scheduling. In *Proceedings of the International Conference on Compilers, Architecture and Synthesis for Embedded Systems (CASES'03)*, pages 255–266. ACM, 2003.

[308] S. Vassiliadis and I. Sourdis. FLUX Networks: Interconnects on Demand. In *Proceedings of the Embedded Computer Systems: Architectures, Modeling and Simulation*, pages 160–167, 2006.

[309] B. Vermeulen, J. Dielissen, K. Goossens, and C. Ciordas. Bringing Communication Networks On Chip: Test and Verification Implications. *IEEE Communications Magazine*, 41(9):74–81, September 2003. Guest editors: Dimitris Gizopoulos and Rob Aitken.

[310] P. C. Vinh. Homomorphism between AOMRC and Hoare Model of Deterministic Reconfiguration Processes in Reconfigurable Computing Systems. *Scientific Annals of Computer Science*, (17):113–145, 2007.

[311] P. C. Vinh. Categorical Approaches to Models and Behaviors of Autonomic Agent Systems. *The International Journal of Cognitive Informatics and Natural Intelligence (IJCiNi)*, 3(1):17–33, January–March 2009.

[312] P. C. Vinh. *Dynamic Reconfigurability in Reconfigurable Computing Systems: Formal Aspects of Computing*. VDM Verlag Dr. Müller, Saarbrücken, Germany, 1st edition, January 2009. 236 pages.

[313] P. C. Vinh. Formal Aspects of Self-* in Autonomic Networked Computing Systems. *In M. Denko, L. Yang and Y. Zhang (Ed.), Autonomic Computing and Networking*, pages 383–412, Springer, 2009.

[314] K. von Arnim, E. Borinski, and P. Seegebrecht. Efficiency of Body Biasing in 90-nm CMOS for Low-Power Digital Circuits. *IEEE Journal of Solid-state Circuits*, 40(7):1549–1556, July 2005.

[315] M. Waldén and K. Sere. Reasoning about Action Systems Using the B-Method. *Formal Methods in System Design*, (13):5–35, 1998. Kluwer Academic Publishers.

[316] D. Wang, H. Matsutani, H. Amano, and M. Koibuchi. A Link Removal Methodology for Networks-on-Chip on Reconfigurable Systems. In *Proceedings of the International Conference on Field Programmable Logic and Application (FPL'08)*, pages 269–274, Heidelberg, Germany, September 8–10, 2008.

[317] H. Wang, L. Peh, and S. Malik. Power-Driven Design of Router Microarchitectures in On-Chip Networks. In *Proceedings of the 36th Annual IEEE/ACM International Symposium on Microarchitecture (MICRO'03)*, pages 105–116, 2003.

[318] L. Wang, Y. Cao, X. Li, and X. Zhu. Application Specific Buffer Allocation for Wormhole Routing Networks-on-Chip. In *Proceedings of MICRO'08*, 2008.

[319] Y. Wang. A Cognitive Informatics Reference Model of Autonomous Agent Systems. *The International Journal of Cognitive Informatics and Natural Intelligence (IJCiNi)*, 3(1):1–16, January–March 2009.

[320] E. W. Weisstein. Heawood Graph. In *MathWorld–A Wolfram Web Resource*, 2010.

[321] D. Wiklund and D. Liu. SoCBUS: Switched Network on Chip for Hard Real Time Systems. In *Proceedings of the International Parallel and Distributed Processing Symposium (IPDPS'03)*, page 78a, 2003.

[322] P. T. Wolkotte, G. J. M. Smit, G. K. Rauwerda, and L. T. Smit. An Energy-Efficient Reconfigurable Circuit-Switched Network-on-Chip. In *Proceedings of the 19th IEEE International Parallel and Distributed Processing Symposium*, page 155, 2005.

[323] F. Worm, P. Ienne, P. Thiran, and G. de Micheli. A Robust Self-Calibrating Transmission Scheme for on-Chip Networks. *IEEE Transactions Very Large Scale Integration Systems*, 13(1):126–139, 2005.

[324] Y. Xie and W. Wolf. Allocation and Scheduling of Conditional Task Graph in Hardware/Software Co-Synthesis. In *Proceedings of the Design Automation Conference (DAC'01)*, 2001.

[325] Q. Xu and J. He. Laws of Parallel Programming with Shared Variables. In D. Till, editor, *Proceedings of the 6th Refinement Workshop*, Workshops in Computing. BCS-FACS, London, Springer-Verlag, 5–7 January 1994.

[326] E. Yahya, O. Elissati, H. Zakaria, L. Fesquet, and M. Renaudin. Programmable/Stoppable Oscillator Based on Self-Timed Rings. In *Proceedings of the 15th IEEE Symposium on Asynchronous Circuits and Systems (ASYNC'09)*, pages 3–12. IEEE Computer Society, 2009.

[327] Y. S. Yang, J. H. Bahn, S. E. Lee, and N. Bagherzadeh. Parallel and Pipeline Processing for Block Cipher Algorithms on a Network-on-Chip. In *Proceedings of the 6th International Conference on Information Technology: New Generations*, pages 849–854, 2009.

[328] T.T. Ye, L. Benini, and G. De Micheli. Analysis of Power Consumption on Switch Fabrics in Network Routers. In *Proceedings of the 39th Design Automation Conference (DAC'02)*, pages 524–529, 2002.

[329] A. Yin, P. Liljeberg, Z. Lu, and H. Tenhunen. Monitoring Agent Based Autonomous Reconfigurable Network-on-Chip. In *Proceedings of the Workshop Digest on Diagnostic Services in Network-on-Chips (DSNoC'08)*, Anaheim, CA, June 8-13, 2008.

[330] A. Younis and Z. Dong. Trends, Features, and Tests of Common and Recently Introduced Global Optimization Methods. *Engineering Optimization*, 42(8):691–718, August 2010.

[331] K. Y. Yun and R. P. Donohue. Pausible Clocking: A First Step toward Heterogeneous Systems. In *Proceedings of the International Conference on Computer Design (ICCD'96)*, pages 118–123. IEEE Computer Society, 1996.

[332] J. Zeppenfeld, A. Bouajila, A. Herkersdorf, and W. Stechele. Towards Scalability and Reliability of Autonomic Systems on Chip. In *Proceedings of the 13th IEEE International Symposium on Object/Component/Service-Oriented Real-Time Distributed Computing Workshops (ISORCW'10)*, pages 73–80, 2010.

[333] Z. Zhang, A. Greiner, and S. Taktak. A Reconfigurable Routing Algorithm for a Fault-Tolerant 2D-Mesh Network-on-Chip. In *Proceedings of the 45th ACM/IEEE on Design Automation Conference (DAC'08)*, pages 441–446, 2008.

[334] W. Zhou, Y. Zhang, and Z. Mao. An Application Specific NoC Mapping for Optimized Delay. In *Proceedings of the IEEE International Conference on Design and Test of Integrated Systems in Nanoscale Technology (DTIS'06)*, pages 184–188, Tunis, Tunisia, September 5–7, 2006.

[335] W. Zhou, Y. Zhang, and Z. Mao. Pareto Based Multi-Objective Mapping IP Cores onto NoC Architectures. In *Proceedings of the IEEE Asia Pacific Conference on Circuits and Systems (APCCAS'06)*, pages 331–334, Singapore, Singapore, December 4–7, 2006.

[336] Y. Zhu and F. Mueller. Feedback Dynamic Voltage Scaling DVS-EDF Scheduling: Correctness and PID-Feedback. In *Proceedings of the Workshop on Compilers and Operating Systems for Low Power*, 2003.

[337] P. Zipf, G. Sassatelli, N. Utlu, N. Saint-Jean, P. Benoit, and M. Glesner. A Decentralised Task Mapping Approach for Homogeneous Multiprocessor Network-On-Chips. *International Journal of Reconfigurable Computing*, 2009.

Index